Preface

WHAT is common to art and science? Creation. Or rather the drive that impels creativity. The thrill of the word and sound, of the colour, lines and shapes of art. The temerity of the scientific hypothesis which extends beyond reality. What is the aim of a creative act in art or science? To surpass reality. Art suggests the infinite variations of reality's manifestations, which are impossible to capture with the usual senses. That such expressions are part of a long and complex chain . . . is all that we know.

One of my teachers at Oxford, a Nobel prizewinner, said: 'We should seek what others have not seen, think what others have not thought of.' Is that not the essence of creation? Malraux lucidly stated this in a text on cultural heritage written in 1936: 'The convincing force of a work . . . lies in the difference between it and the works that preceded it.' He illustrated the subject by quoting Giotto, but could have made his point just as well by discussing Einstein's theory of relativity.

Occasionally, when science reaches beyond its frontiers, it merges with philosophy. Likewise, art can be dematerialized – boiled down to pure ideas. Artists exercise the same self-discipline and rigour as scientists.

Creation, whether in art or science, is a long journey. Some believe that youth is a prerequisite of creativity. This is not necessarily the case. We obviously all admire Mozart's precocity, but equally admirable is the expression of a maturer mind, one whose critical faculties have been nurtured over time and through experience.

It is difficult to be thrilled by anything that is too neatly served up. I do not particularly appreciate pure evidence, creations to which nothing can be added. I much prefer to come upon works in the making, which draw audiences into the exhilarating struggle of creation – in which anyone can be a co-creator, a participator in the act of creation. This is what I thought recently while listening to the great cellist Rostropovich who, at every performance, recreates music with unparalleled enthusiasm: it is this invitation to share beauty that embodies the true act of genius.

'Dare to know': such was the motto at Oxford. Perhaps the opposite is even more true: 'Know how to dare.' To dare to invent, to innovate and create, to escape routine and provoke the unpredictable. As the days roll on, until the very end, we should fully reinvent each day itself and dare to paint it with fresh colours.

Federico Mayor
Director-General of UNESCO

Art and Science

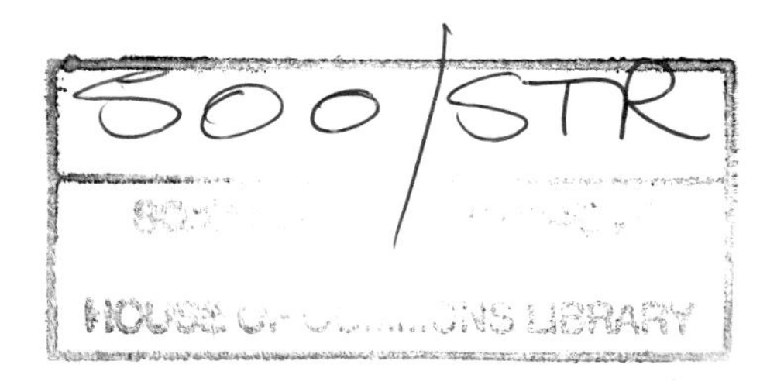

Art and Science

Eliane Strosberg

UNESCO PUBLISHING

The author is responsible for the choice and the presentation of the facts contained in this book and for the opinions expressed therein, which are not necessarily those of UNESCO and do not commit the Organization.
The designations employed and the presentation of material throughout this publication do not imply the expression of any opinion whatsoever on the part of UNESCO concerning the legal status of any country, territory, city or area of its authorities, or concerning the delimitation of its frontiers or boundaries.

Published in 1999 by the United Nations Educational, Scientific and Cultural Organization

7, place de Fontenoy, 75352 Paris 07 SP, France

Design by Gérard Prosper
Cover by Jean-Francis Chériez
Printing by Groupe Corlet - N° 40309

ISBN 92-3-103502-9

Foreword

This book is part of a UNESCO transdisciplinary cultural programme. Its style is deliberately casual and narrative; it is not a history book, but rather a collection of stories aimed at stimulating interest in the subject through dialogue.

The project deals predominantly with the visual arts. The first part, on architecture, is a general overview. Decoration, painting and graphic communication constitute the other main sections. Music and the performing arts are only dealt with in the context of their relationship with the visual arts. Literature is not included in this work. Given the vastness of the subject, the main focus is on Western culture, although numerous examples from other civilizations have been included.

A number of eminent personalities have supported this effort: René Berger, writer and honorary president of the arts department at the University of Lausanne; Jean Dausset, Nobel laureate in medicine and modern art connoisseur; Frank Popper, specialist in philosophical questions relating to art and technology; Ilya Prigogine, Nobel laureate in chemistry and author of books on aesthetics.

The text also received input from specialists in their respective fields: Marianne Clouzot, artist; Philippe Comar, artist and professor at the École Supérieure des Beaux-Arts; Françoise Gaillard, professor of philosophy at the University of Paris VII; Antoinette Hallé, curator of the Musée de la Céramique at Sèvres; Bruno Jacomy, engineer, associate director of the Musée des Arts et Métiers; Elaine Koss, deputy director of the Art College Association; Bernard Maitte, specialist in the history of science and knowledge; Sarah McFadden, art historian and an editor of *Art in America.*

This book would not have been possible without the generosity of those who lent their iconographic documents as a courtesy – artists, photographers, museums, librarians, publishers and corporations.

The author would like to thank Tereza Wagner and Michiko Tanaka who literally carried this project within UNESCO. She is immensely grateful to Donny, Serge and Muriel for their unconditional support, and wishes to thank all those who contributed directly or indirectly to realize this project: Isi and Lilly Leuwenkroon, Jules and Julia Strosberg, Robert and Evelyn Leuwenkroon, Mouche Strosberg and Vera Bakhoven, Ralph, Hetty and Dan Lopes Cardozo, Rosette and Christian Wuilmart, Luc and Soraya Leuwenkroon, Françoise Russo-Marie, Jean-Pierre Mégnin, Caroline Lawrence, Ellen Levy and Elsa Papatechera. With a special thought for Ed Haber and Bill Mandy.

Contents

The art and science dialogue

While the interaction between artists and scientists is often fruitful, a true dialogue is not always easy to bring about. To begin with, dictionaries offer several definitions of 'art'. One describes art as a form of 'science or knowledge'. Another suggests that 'art is a series of means and procedures tending towards an end'. In some dictionaries, the concept of beauty appears in only fifth position as an element in art. Needless to say, creators do not depend on such descriptions to define who they are and what they do.

Many consider that works of art should be appreciated for their intrinsic value or their innovative vision of the world. In the past, art served religion, magnified the power of patrons, reflected skills aiming at producing elegant objects. Nowadays it is used for self-expression, and even as therapy.

When it comes to the word 'science', most dictionaries offer a description which seems, at first glance, quite obvious. Science is the knowledge of the laws of nature. In other words, it embodies all studies which carry a universal meaning and which are pursued by research methods based on objective and verifiable facts.

However, we should bear in mind that for centuries, metaphysics, theology and philosophy prevailed; science, too, was once nurtured by beliefs. Concepts such as 'method and objectivity' have appeared only recently, and to some, science still remains mysterious.

MONA-LEO, LILLIAN SCHWARTZ, 1986
Mona-Leo is a combination of computer-juxtaposed image. The result suggests a strange symmetry between the face of Mona Lisa and that of Leonardo da Vinci.

Lillian and Laurens Schwartz, The Computer Artist's Handbook. W. W. Norton, Inc., 1992

Divergence and convergence

According to the mathematician and philosopher, Bertrand Russell: 'In art, nothing worth doing can be done without genius; in science, even a moderate capacity can contribute to supreme achievement.' Such a strong opinion deserves a few comments.

Whereas the artist tries to stir emotions, the scientist has to convince. Art looks into the 'why'; science also raises the question of 'how'. For the Cubist painter Georges Braque: 'art provokes, while science tries to reassure.' Science, working towards collectively recognized and precise objectives, tries to remove ambiguities, which art accepts and even emphasizes as inevitable in the realm of subjective experience.

It is commonly thought that everyone has the ability to appreciate art, while science is accessible only to some. What is more, scientists and artists generally consider themselves very different from each other. A recent 'left brain/right brain' hypothesis reinforces this notion. It states that scientists, whose tasks are primarily logical and analytical, mainly use the left side of the brain; the right side, seat of intuition and imagination, would be more developed in the artist.

Despite, or perhaps because of their divergence, artists and scientists are bound by a mutual fascination: opposites attract. Is it that artists' need to draw from science just expresses the urge to use whatever means available, to serve their art? Maybe scientists' search for convergence simply stems from an inclination to create coherent models to explain the world.

Together, art and science conquer innovating concepts, often using the same subjects to the same end. Giving birth to ideas and forms is what makes an artist or a scientist. To scrutinize the cosmos, examine nature or study the brain, are explorations common to both. Following parallel paths, art and science are in many ways mutually enabling.

Cubist painting, for example, seemed to presage the theory of relativity (whose delay may have been due to the complexity of the language required for scientific demonstration). In architecture and the performing arts, however, science and technology often work as catalysts.

'Who is ahead of whom' is irrelevant, because the other always catches up in the end. Paul Valéry, a French writer, thought that: 'Science and art are crude names, in rough opposition. To be true, they are inseparable . . . I cannot clearly see the differences between the two, being placed naturally in a situation where I deal only with works reflecting thinking matters.'

SYMMETRY PATTERNS FOUND IN DIFFERENT CULTURES, PETER STEVENS, 1981
During Antiquity, artists produced these schemes in their decorations. Not until the nineteenth century were mathematicians able to analyse and duplicate such motifs.
Handbook of Regular Patterns. *The MIT Press*

Aesthetics and method

The word 'aesthetics' designates a branch of philosophy concerned with the 'science of beauty' in nature and art. Beauty and discipline are important for both artists and scientists. The latter readily admit that logical reasoning is sometimes overestimated, and acknowledge that the use of imagination is an integral part of their creative process.

Scientific theories may take years to develop, during which time aesthetic consideration plays a major role. 'A beautiful theory killed by a nasty little fact' said Thomas Henry Huxley, a biologist and science popularizer. Many scientists find their greatest satisfaction in aesthetic contemplation and describe their research as a quest for beauty. Masterpieces such as Aristotle's *Physics* or Newton's *Optics*, seduce first and foremost by the elegance of their logic.

Some of the most famous artists, on the other hand, frequently place discipline and method above aesthetic consideration. Bach, a brilliant manipulator of 'ready-made formulas', considered himself a craftsman: 'I have to work very hard; whoever works as hard will get as far.' Beauty did not seem to be the only goal of his music, composed from 'divine mathematics'. Igor Stravinsky's approach to music was extremely laborious and structured, too. He portrayed the musician as 'a craftsman whose materials of pitch and rhythm in themselves harbour no more expression than the carpenter's beam or the jeweller's stone'.

Painters can be equally disciplined. Thus this remark by the post-Impressionist Seurat: 'To see poetry in what I have done . . . No, I applied my method and that's all there is to it.' As for Matisse's preoccupations with his huge compositions, they seem to be matched by the mental gymnastics of scientists deeply absorbed in shaping their hypotheses. Guided by formal rigour, he stated: 'A moment comes when each part has found its legitimate relationship, and from there on, it would be impossible to add a single line to the image without having to start the painting all over again.' Georges Braque even said: 'I love the rule which corrects emotion.'

Although science aspires to objectivity, scientists are not always more objective than artists. Galileo deliberately ignored Kepler's work, and Pasteur did not immediately adhere to Darwin's. The need to test established knowledge, the thrill of exploring a new domain, the hope of discovering a new order, all compel scientists to accept the same basic challenge: to reveal what no one has expressed before, and thus leave their mark.

Most creators are totally committed to their tasks. Whatever the nature of their skills, those who succeed in imposing an innovative vision of the world often enjoy particularly productive careers.

Polyvalence

Despite their commitment to their chosen field, artists such as Leonardo da Vinci and Albrecht Dürer were also gifted scientists. Conversely, scientists like Nikolaus Copernicus and Louis Pasteur were talented artists. Such relations are by no means limited to the visual arts. The number of mathematician-musicians (for example Euler, Schweitzer and Einstein) is even more striking.

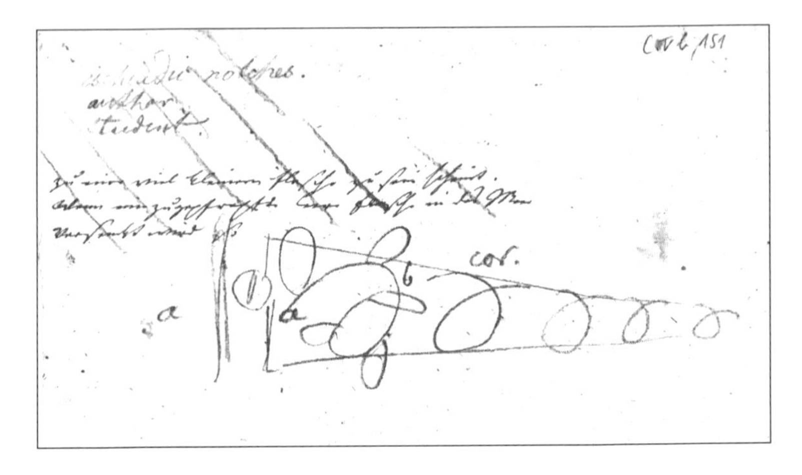

The Spiral Forms in Nature,
Johann Wolfgang von Goethe, 1831
The Romantic writer, Goethe, was also a biologist. He introduced the notion of 'morphology' which would appear to be fundamental to the development of the theory of evolution of species. This drawing suggests the mutation of a leaf into a flower.
Stiftung Weimarer Klassik. Goethe-Schiller-Archiv. Photo S. Geske

Numerous artists have been interested in science, and scientists in art, but most of them focus on only one domain, while their interest in the other remains subsidiary. Time constraints and technical barriers suffice to explain such a dissociation. Moreover, one cannot underestimate the distinction between interest and real talent (interest alone by no means leads to significant performance).

Can one easily switch from art to science and vice versa? For an artist to become a scientist is placing the stakes very high, but the reverse is no easier. What accounts for creativity and who is destined to excel in its pursuit? Such questions have long puzzled great minds. According to the Dutch chemist, Jacob van t'Hoff, who was the first Nobel prizewinner in chemistry: 'The most innovative scientists are almost always artists, musicians or poets.' Creativity might indeed depend on a capacity to integrate traditionally incompatible forms of experience. But this is not always the case: Darwin or Cézanne, to cite just two examples, were not particularly polyvalent.

Still the question remains: what favours, in some of the most inventive minds, an aptitude and appetite for both science and art? Certain forms of art may exhibit affinities to scientific disciplines – and hence mental processes may link the architect to the astronomer, the stage director to the physicist, the psychologist to the painter. Perhaps a chemist thinks like a decorator, and a mathematician like a musician.

Inspiration and visualization

Centuries ago, creators were expected to follow in the footsteps of their predecessors, building along a prescribed tradition. Nowadays, innovation has become the driving force in art, as the public is increasingly demanding. Yet novelty, too, whether in art or science, requires a base. (For example, examination of Picasso's painting *Les Demoiselles d'Avignon* reveals neo-classical sources [Delacroix and Ingres] as well as less familiar ones of African and Polynesian statuary.) The stylistic sources behind a work of art are often heterogeneous and more difficult to trace than are the origins of scientific theory; moreover, the artist is not required to explain them.

Since creation can never be fundamentally new, there might be no such thing as a muse or a mystery of creativity, but various interpretations co-exist. The 'unconscious' eventually replaced the role of the gods as the creative source. In the nineteenth century, Henri Poincaré, a mathematician and philosopher, advanced a theory according to which the creative process passes through various stages such as preparation, incubation, illumination, verification. In this way, numerous thought combinations would be tested by the 'unconscious', and only those meeting some sense of harmony would be selected.

Encounters with an idea or an image are essential in art and science. In order to create, both need to be visualized either mentally or on paper, or through measurement. Quantification and visualization may be said to represent two sides of the same coin.

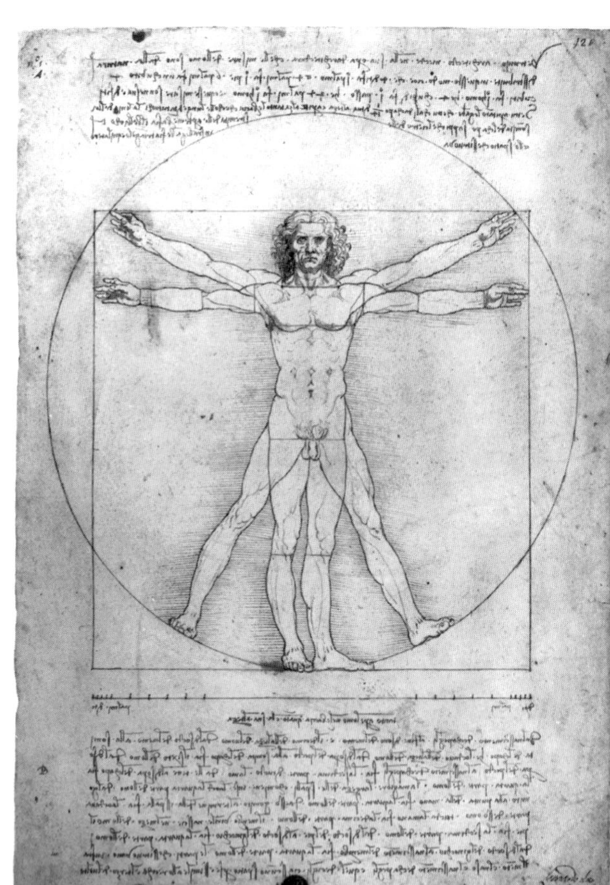

The Vitruvian Man, Leonardo da Vinci, 1492
During Antiquity, human proportions were used to determine the dimensions of sculptures and monuments. This practice re-emerged in the Renaissance and is still in use. The circle and the square in which *The Vitruvian Man* is inscribed symbolized, respectively, the cosmos and the earth – analogies of the macro and micro cosmos, with man at the centre of the universe.
Gallerie dell'Accademia, Venice. Ministerio per i Beni Culturali e Ambientali

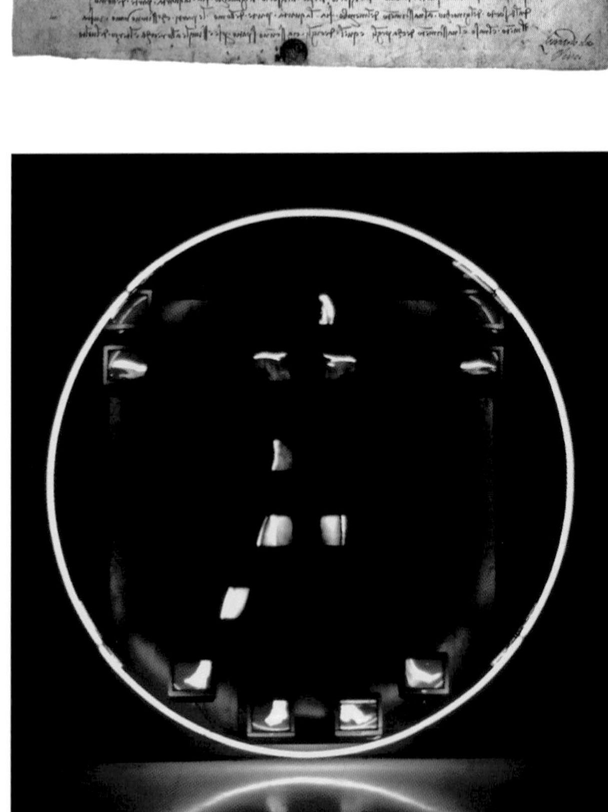

Fragments of an Archetype, Catherine Ikam, 1980
This monumental sculpture was one of a series of installations. Composed of sixteen video recorders, it was placed inside a huge neon circle.
Displayed in 1980 at the Centre Georges-Pompidou, Paris

The physicist Wolfgang Pauli studied the role of symbolism through collective models applied to scientific concepts. Many artists and inventors linked their discoveries to epiphanic moments when their minds were 'floating', describing these through poetic images which often result from legends; yet, because of their recurrence – think of 'eureka', Archimedes' cry in his bath – they deserve our attention.

What should one think of Newton's sudden revelation while watching an apple falling from a tree? Gutenberg reportedly said that his idea for the first printing press struck him like a ray of light, while he was observing a wine press operating at a festival. Lumière said that he invented the moving picture system while watching his mother using the sewing machine. Einstein declared that latent mental images stimulated his imagination for years, before he was able to draw any conclusions from them: 'In my case (psychical entities) are of a visual and sometimes of a muscular type. Conventional words and other

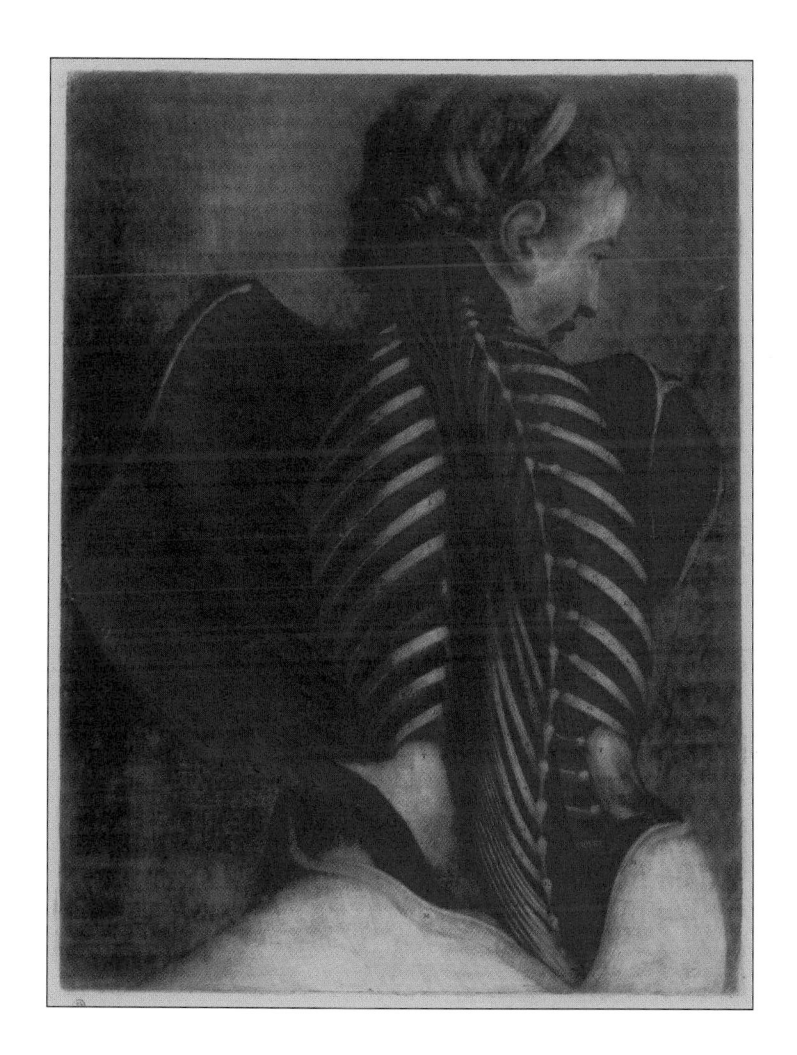

THE ANATOMICAL ANGEL, JACQUES GAUTIER D'AGOTY, EIGHTEENTH CENTURY

Artists, like scientists, use visualization methods. Inspired by Newton's *Optics*, Jacob Christophe le Blon invented a technique (1710) for colour engraving based on the combination of the three primary colours: yellow, red and blue. Through a purely mental exercise, the optical mixture had to be conceived for each spot of the image.

Bibliothèque Nationale de France. Photo Bibliothèque Nationale de France

signs have to be searched for laboriously only in a second phase.'

Some of the world's greatest scientists – Aristotle, Alhazen, Bacon, Descartes, Einstein – were fascinated by optics and visualization. So, too, some of the greatest painters – Velázquez, Vermeer and Turner – methodically analysed light, colour and image formation. And did not Titian or Bonnard paint most radiant works after they had almost lost their sight? 'What I did not draw, I did not see,' said Goethe.

In the past, researchers were puzzled by the mysterious functioning of the eye. Today, they try to understand how the brain decodes the images it receives from the eye.

Studies of twentieth-century geniuses – Picasso, Einstein, Freud and Stravinsky – conclude that no universal characteristic has yet been found to link their phenomenal creativity, other than the spirit of independence. What makes a genius remains a riddle worthy of the Sphinx.

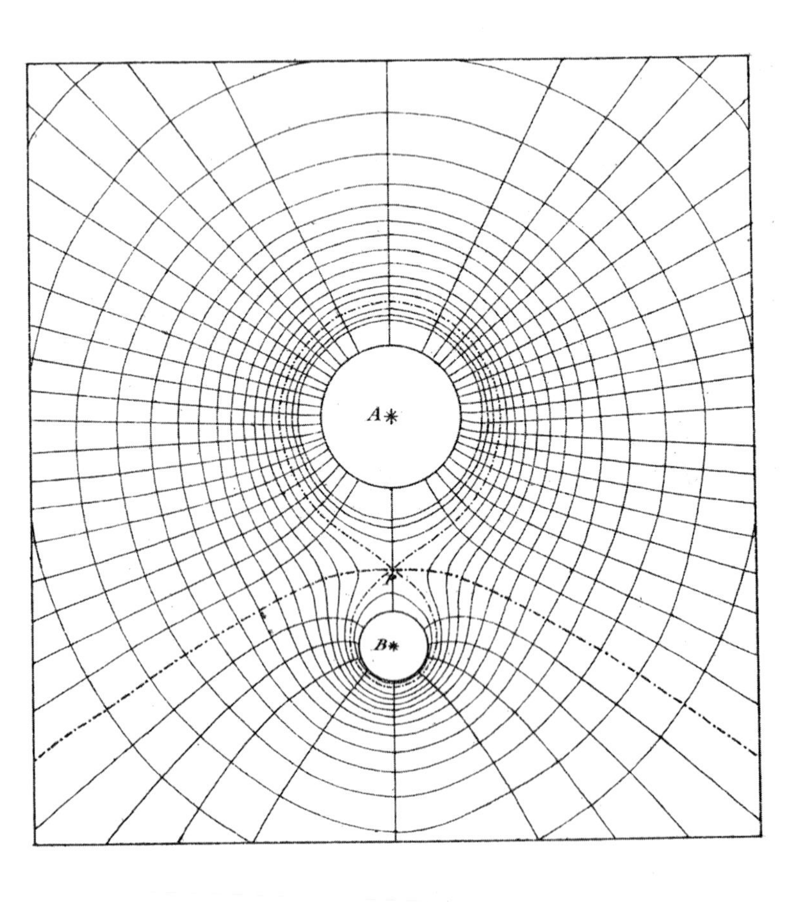

Electromagnetic field lines, James Clerk Maxwell, 1871

The physicist Maxwell visualized the concept of an electrical field, graphically, through 'experimental lines of forces on paper'. This notion of a space capable of transmitting radiation emerged along with progress in electricity and magnetism. Fascinated by optical phenomena, Maxwell was also an important player in the development of photography, and produced the first colour photograph.

QC 518 M46. Library of Congress, Washington, D.C.

Research

Some artists do not sketch: they let the painting reveal itself as they go along. Others proceed through trial, error and corrections, as do scientists. In this manner, Cézanne explored the effect of light on his *Montagne Saint-Victoire* – a theme to which he returned, over and over again. Matisse's letters about his own privileged theme, *La Danse,* reveal a similar concern: to solve a problem. Changing the place of a single mark or slightly modifying a colour in a painting can easily destroy the overall effect.

When analysing Picasso's preliminary drawings for major paintings or collages, sequential steps in his process of constructing a picture can be retraced. Calder, another creator who enjoyed a long career, allows us to trace his progressive approach in sculpture. Newton and Einstein reworked their equations and mental images during long years of continuous self-training. The excitement generated by enigmas (like jigsaw and crossword puzzles) seems to develop creativity.

When a concept no longer holds up under critical scrutiny, analysis of the inadequacies calls for modifications. Any new concept constitutes a challenge to the existing ones, which must be surpassed. Fundamental changes rarely occur in a flash, but tend to evolve as the difficulties are being solved, one after another.

To search . . . up to what point? According to the physicist Pauli: 'The happiness that man feels in understanding seems to be based on inner images pre-existent in the human psyche with the external objects and their behaviour.' While Braque noted that: 'The painting is finished when it has erased the idea.'

According to Pierre Teilhard de Chardin, a Jesuit palaeontologist: 'The spirit of research and conquest is the supreme essence of evolution. It penetrates all those who have ever dedicated their intelligence and their lives to art or science.'

Education and new technologies

EXHIBITION VIEW, JOSEPH BEUYS
The German artist Beuys proclaimed that 'everybody is an artist' and everything is art, including political and social processes. He used the most ordinary objects and materials in his performances, and displayed didactic skills to expound on his philosophy for a new democracy.

Nowadays, specializing has become inevitable, but over-specializing destroys creativity, which thrives on open-mindedness. Play remains a primary form of learning, whatever the age.

Goethe and Edison never had a formal education. Einstein and Picasso opposed the traditional school system, and other creators ignored academic training. It is thus tempting to assume that having no constraints provides a better chance of approaching the world creatively. However, the value of a formative but flexible training can never be understated.

Education, expanding beyond traditional segmentation, is being redefined. Psychologists suggest the existence of transversal forms of intelligence: spatial, social, verbal, etc., which would go beyond traditional barriers of artistic or scientific intelligence.

Will researchers identify 'creativity genes'? Creativity is surely not entirely determined by genetic heritage; in many ways, the social and cultural environment remains a key factor.

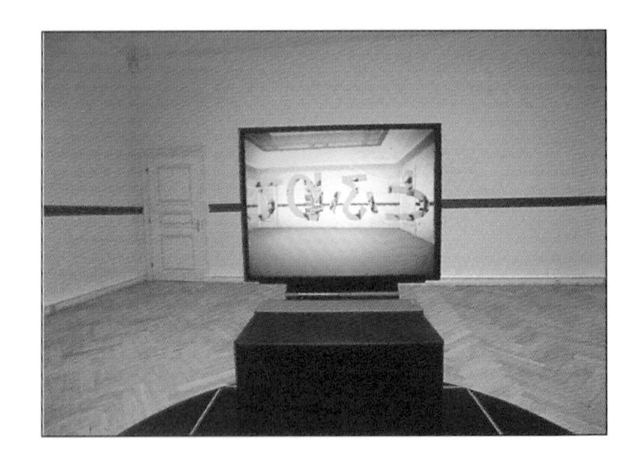

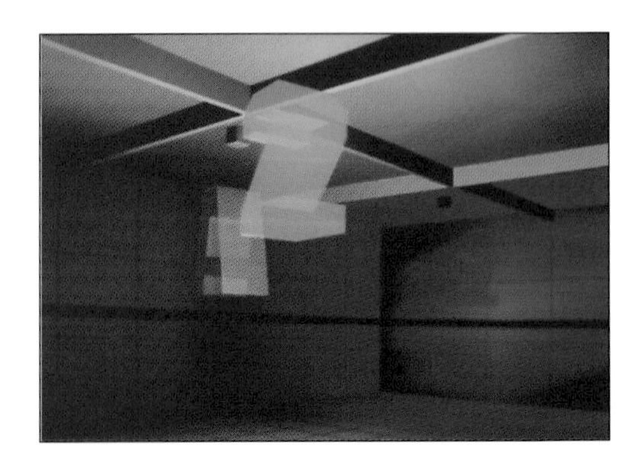

VIRTUAL MUSEUM, JEFFREY SHAW, 1991
Inside this interactive work of art, one can explore different rooms of an imaginary museum containing paintings, sculptures and computer-designed images.

Artists, like scientists, are increasingly drawn into multidisciplinary activities because of economic pressure, and sometimes by choice.

Art and science use the same tools and materials; technique then becomes their main link. According to Marshall McLuhan, the visionary media specialist: 'the medium is the message.' Convergence is particularly strong during mutations, as is the case now. Interestingly enough, technology is a recent term designating the industrial applications of science. (In daily language, the words 'technique' and 'technology' are often used interchangeably, even though technology is considered to be 'nobler', because it is thought to be more scientific.)

In a computerized and digitalized world, the spectator becomes a permanent student – whether willingly or not. The writer Arthur Koestler thought that: 'Creativity is a type of learning process where the teacher and the pupil are located in the same individual.' 'Creativity', 'education', 'entertainment' are now intricately woven concepts because new technologies favour the relationship between artist, creation and spectator.

This link is considerably reinforced through modern means of communication, potentially transforming social order altogether. Of course, technology will not bring all the answers. But, thanks to new media, it is indeed conceivable that everyone has a creative potential that can be explored, in art and science, or any other domain.

ORB, BILL PARKER, 1990
Work of art, gadget or scientific tool? This glass sphere containing a combination of rare gases reacts to heat upon touching, by emitting light through an electrode at its centre. To observe, play and learn was this creator's intention.
Photo Julie Walker

A dynamic history

Antoine Laurent de Lavoisier and His Wife, Jacques Louis David, 1788
Lavoisier was the father of the chemical revolution. His wife was an artist who studied with David, painter of the above portrait. She also illustrated Lavoisier's laboratory notes and translated English scientific papers.

The artist-scientist

Today, most artists and scientists live in distinct socio-economic worlds, but this was not always the case. It is of course impossible to know if the Stone-Age artist was considered equal to the healer, although the abundance of expertly executed cave paintings suggests that artists were already highly skilled in the art of representation. The mere fact that they were afforded the time and means to practise during the daily struggle for survival, allows us to hypothesize that the painter's function was important.

A clue to the artist's social status in ancient times is contained in the Hammurabi code (*c.* 1750 B.C.), one of the oldest written law codes. The architect and the sculptor were considered equals to the butcher and the metal worker, whose functions were closely associated with ritual practices. Knowledge and art – architecture, sculpture and painting – came under the authority of priests. The major task of artists, whose names were not recorded, was to master materials in a strictly prescribed manner.

Creations bearing a signature first appeared in Ancient Greece; however, creativity in the contemporary sense was foreign to the Greeks, for whom arts and crafts were synonymous. The motivating force of self-expression was referred to as *techne*.

GREEK VASE PAINTING REPRESENTING A FOUNDRY, FIFTH CENTURY B.C.

The Greek god Hephaestus (Vulcan to the Romans) is said to have wrested fire from the earth's bowels and used it to smelt metal. The union of the Prime Metallurgist with Aphrodite (Venus) symbolized the alliance of science with beauty. In Ancient Greece *techne* inspired art; the Greek verb *tikein* means to create.

Staatliche Museen zu Berlin. © bpk, Photo J. Laurentius

Aristotle wrote: 'The nature of all *techne*, is to understand the genesis of a work of art, to research the technique and theory behind it, to find its principles in the person who gave birth to it, and not in the creation itself.' Originally, 'technique' did not simply designate methods of fabrication; it carried symbolic and spiritual connotations.

The Romans copied their Greek predecessors and it is often thought that they excelled in feats of engineering rather than art. After the fall of Rome in the fifth century, interest in both art and science faded and their memory was obliterated by repeated invasions. Knowledge re-emerged slowly. During the Middle Ages, the building of cathedrals and the growth of cities allowed the artisans to gain their independence.

In the fifteenth century, Lorenzo de Medici (1449–92) founded a school for the most gifted in order to provide them with basic training in geometry, grammar, philosophy and history. The success of this organization was such that, within a few decades, more than 1,000 academies had sprung up in Italy alone. Within these institutions, a grounding in art meant learning the rules of perspective and studying anatomy, just as it does in art schools today.

Renaissance artisans gradually gained respect and repute. As interest in their work increased, so did the attention paid to the individual artist. Filippo Brunelleschi (1377–1446), the developer of one-point perspective and designer of the famous Florence Cathedral dome, was the first to bear the title of

PYTHAGORAS, CHARTRES CATHEDRAL,
LATE TWELFTH CENTURY

During the Middle Ages an important school of philosophy was established at Chartres where subjects such as optics were studied. This sculpture of Pythagoras is located near the musical instruments. Aristotle (not visible on this image) is represented with the symbols of dialectics; Euclid is accompanied by the geometer's tools, and Ptolemy by the astronomer's.

PAINTING ACADEMY IN ROME,
PIERFRANCESCO ALBERTI, SIXTEENTH CENTURY

Brunelleschi and Botticelli made decorative as well as functional objects. The academies functioned like modern research laboratories, with specialists working under the supervision of a co-ordinating director. Left, a human corpse is dissected.

architect. Just one century later, Michelangelo (1475–1564) would be described as a 'genius' by the public.

While Renaissance artists dared to challenge popes, most of their scientific contemporaries had to pursue research, secretly, and were viewed as enlightened amateurs. Yet they used similar measuring instruments to those of artists, in order to respond to analogous preoccupations. In fact, artists and scientists functioned in parallel.

By opposing a logical explanation of the world to spiritual ones, science would slowly erode religious dogma. Newton, the author of the theory of universal gravitation, became immensely famous while still alive; but most scientists would not benefit from any official status until the nineteenth century.

Isaac Newton, William Blake, eighteenth century
William Blake, the painter, engraver and visionary poet who made this ironic portrait, was a precursor of the Romantic artists.
Tate Gallery, London. Tate Pictury Library, 1998

The rupture in modern times

To explain the workings of the universe, scientists increasingly favoured mechanical concepts, leaving emotional concerns to the purview of art. The more science progressed, the more artists rebelled against it. While the French Revolution destroyed artistic patronage, it spurred technical inventions such as printing and photography. The illustrator, until then an essential recorder of historic moments, suddenly felt threatened. His task could seemingly be carried out just as well, if not better, by machines!

The profusion of scientific inventions in the early nineteenth century gave rise to new terminology. The word 'scientist', coined by the British in 1863, and constructed according to the same logic as the word 'artist', slowly replaced the traditional designation 'natural philosopher', still in use.

Scientific developments were heavily dependent on industry and economics, and thus inextricably linked to power. In response to the public's craving for knowledge, popularizing magazines such as *Scientific American* began publication. From then on, science would play a central role in society and researchers finally became paid professionals.

In the meantime, since artists were no longer needed to represent reality, many took upon themselves the challenge of interpreting it. In so doing, they unwittingly re-established conceptual links with science. Despite their role as 'society's conscience', few artists today have attained the social status of scientists. For example, there is no Nobel prize in the visual arts.

Artists' rewards are often aleatory, whereas scientists generally receive salaries. Of course, the market occasionally propels an artist to unimaginable heights of fame and fortune. Although artists' earnings are generally meagre, they enjoy a precious privilege: relative freedom in creation.

The role of museums

Throughout history, works of art have been made that resemble true marvels of science and technology. Sometimes, it is impossible to determine whether a creation is of the artistic or scientific order. Understanding an object's significance contributes to the emotion it induces.

Art can be enjoyed at various levels; aesthetic landmarks exist in science as well, but their appreciation is generally left to specialists. The teaching of science is not expected to emphasize aesthetic aspects and tends to concentrate on theories that 'work'.

Traditionally, art's end point was 'to produce beauty', not to provide information. However, art is a testimony and, as such, generates knowledge. The research of contemporary artists often resembles a mental construction, transmitted in a visual form, appealing more to the intellect than to the senses. Artists originate ideas which, when sufficiently innovative, become a form of information.

Museums are reliable indicators of the value that a society places on culture. Created in Ancient Greece around 300 B.C., the Museion was a sanctuary for the muses, a study centre principally for mathematics – although specimens of plants and animals were assembled there. During the Renaissance,

Gallery of the Louvre,
Samuel F. B. Morse, 1831–33
As of 1737, the Royal Academy of Painting and Sculpture started to organize regular exhibitions at the Louvre. Artists could hone their skills by copying the masters, but the general public was rarely admitted. Samuel Morse (1791–1872) who developed the electrical telegraph and its code, also painted at the Louvre. He was president of the New York Academy of Design.
Oil on canvas, 73¾" × 108". Terra Foundation for the Arts. Daniel J. Terra Collection, 1922.51. Photograph courtesy of Terra Museum of American Art, Chicago

'antiquarians' started once again to collect objects which were displayed in their 'curiosity cabinets'.

In the eighteenth century, the age of the great explorers, the idea of a national museum took root in various countries, as it became increasingly necessary to house nature's marvels as well as multiple treasures of the wealthy. The next step might possibly have been the creation of a vast 'encyclopedic space', but for practical reasons, specialized museums were favoured instead. Science and art had taken divergent routes.

Regardless of the nature of their collections, museums today give priority to activities such as conservation, research and teaching. In art museums, vast spaces lavishly deploy technology to enhance the presentation of paintings and sculpture. Science museums exhibit their collections with an aesthetic vantage point. In this way, museums, ancient or modern, remind us that art, science and spirituality are intrinsically linked. As knowledge repositories, they are real teaching laboratories: 'Museums are houses which only host thoughts,' said Marcel Proust.

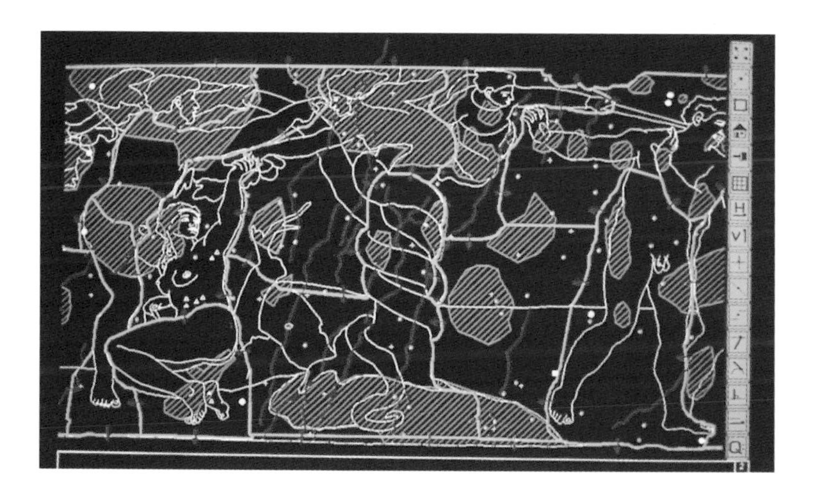

RESTORATION OF THE SISTINE CHAPEL
This computerized map illustrates the restoration process. Conservation has always been a subject of intense controversy. In Imperial Russia, hundreds of paintings were detached from their original supports with deleterious effects. The contemporary approach emphasizes damage prevention rather than repair – in the manner of modern medical practices, and utilizing some of its tools: X-rays, lasers, electronic microscopes, echography.

Art and science: an adventure

Modern art history starts in the Renaissance, when patrons such as the Medici family began collecting Greco-Roman objects. It did not unfold chronologically, but rather like a puzzle – whose different pieces appeared at irregular intervals, in places without any apparent link.

Another two centuries elapsed before scientists started to show interest in prehistory. When cave art was discovered, scholars were at a loss to determine its origin. Indeed, how could they have imagined that such astonishing works were made prior to biblical times? (The term 'prehistory' was coined in 1867.)

The history of art varied through time and from one culture to another. Sculpture, for instance, was considered a major art form in the Western world, whereas in Ancient China it was regarded as just another product of manual labour carried out by the lower classes. Conversely, in the West writing was viewed as a communication tool, while in China, it was revered as the highest art.

For a long time, the story of art was one of styles. New currents were received with scepticism. The Romanesque style was named for its resemblance to Roman art, considered unrefined. Gothic, from the word 'Goth', was brought into use by Renaissance humanists, with pejorative intent. Arrogance was expressed towards those whom critics dubbed 'Impressionists' or 'Fauves'! (In science, too, innovation can disturb; concepts such as 'universal gravitation' or the Big Bang were long subject to scientists' scornful laughter.)

Initially developed by amateurs, art history has become a discipline using advanced technology: measurement of carbon-14 levels in organic material reveals the age of ancient objects (this technique was invented in 1946 by Willard Frank Libby, whose combined passion for chemistry and archaeology earned him the Nobel prize). The genetic study of ancient populations, and of organic materials (parchment manuscripts, bone artefacts, etc.) have prompted a complete reinterpretation of many concepts. Revolutionary results sent historians of all kinds back to their books.

Art history's interpretation, or questions of conservation, are influenced by multiple factors: biography, stylistic analysis, iconography, psychology, socio-political analysis, feminism,

THE BEGINNINGS OF ART HISTORY,
SERGE STROSBERG, 1998
Destroyed by Vesuvius' eruption, the town of Pompeii was rediscovered in the eighteenth century. The information gathered about Ancient Roman culture had a considerable impact on the development of art history. The excavations were of such importance that the royal family kept the operation secret. Even Johann Winckelmann, the father of 'scientific' archaeology, was admitted with great difficulty to the site. Later, when European nobility flocked to Italy, 'surprise excavations' with 'guaranteed discoveries' were organized to that effect!

structuralism, and so on. . . . Science now plays a key role, but many see herein yet another trendy sign. Moreover, science is not infallible . . . its history has fluctuated, like that of art.

During Antiquity, science would alternately go back to its technical or spiritual roots: for example, in Greece *c.* 500 B.C., *sophia* meant 'technical ability' before it became associated with the notion of wisdom.

Several schools of thought functioned in parallel: there were the Materialists, while Plato's followers took a conceptual approach; nature observers influenced by Aristotle; and finally the pragmatic thinkers of Alexandria. Philosophers revised rival theories, establishing in this way what would much later be called 'an intellectual tradition'.

Scientia, in medieval Latin, referred to knowledge in general; reasoning, although based on different assumptions from ours, was nevertheless often elegant. The Arabs practised, commented, taught and transmitted science.

CURIOSITY CABINET, SERGE STROSBERG, 1998
During the Renaissance, collectors assembled coins, instruments, fossils, anatomical specimens, and miscellaneous objects. In 1657, Leopoldo de Medici founded the *Academia del Cimento* – the first organized research institute since the destruction of the museum of Alexandria in 641. Modern research started with basic activities of assembling, comparing and classifying.

Through an approach which combined technical and intellectual exercises, research in the West was reactivated during the Renaissance, paving the way for new fields of investigation. A progressive thought pattern slowly emerged.

Research methods varied from one discipline to another. Eighteenth-century physics was quantitative and deductive; nineteenth-century biology – for example, Darwin's theory of evolution – was qualitative. Despite their common search for truth, the many domains of science proceed at diverse paces, employing different practices and models.

Scientific innovation is scarcely conceivable without the assimilation of prior knowledge; established facts need to be re-evaluated. There are no specific criteria defining a breakthrough, that is, before experimental proof. Research picks up speed when different groups tacitly adhere to a scheme, as happens in art. And changes often take place in the midst of competition, passion and anxiety!

Science is no longer synonymous with a quest for the absolute truth, as recently developed theories of 'chaos' and 'uncertainty' set an inherent limit to knowledge itself. Scientific method divides reality into segments, which permits the examination of tiny cross-sections. This reduction is the great enabler of scientific success. All sciences, whether descriptive like botany, or structural like physics and mathematics, are contained within a 'frame' which allows scientists to create a logical order across a multitude of phenomena.

Science does not always follow a linear path. Breaks occur, such as with Newton's physics and Einstein's theory of relativity. Alexandre Koyré, a modern historian, considered that: 'history is not the reverse progress of science, that is the study of outdated steps whose modern truth would be the vanishing point. It should, on the contrary, be an effort to research and explain to what extent ancient attitudes surpass previous notions in their own day.'

While developments in science can, by definition, quickly become obsolete, art at its best presumably never ages. Picasso said: 'To me, there is no past or future in art. If a work of art cannot live always in the present, it must not be considered at all. The art of the Greeks, of the Egyptians, of the great painters who lived in other times, is not an art of the past; perhaps it is more alive today than it ever was.' In art, as in science, content might age, even though the aesthetic dimension

Relativity, Maurits Cornelis Escher, 1953

This lithograph by the Dutch artist Escher presents a visual paradox, combining three distinct perspective views into a coherent whole. According to Arthur Koestler: 'Einstein's space is no closer to reality than van Gogh's sky. The glory of science is not in a truth more absolute than the truth of Bach or Tolstoy, but in the act of creation itself. The scientist's discoveries impose his own order on chaos, as the composer or painter imposes his; an order that always refers to limited aspects of reality, and is based on the observer's frame of reference, which differs from period to period as a Rembrandt nude differs from a nude by Manet.'

remains. Being implicitly bound to the evolution of human thought, knowledge tends to develop in several domains simultaneously.

Taking, for example, the 'evolution' of the pictorial style from Manet to Cézanne, some might consider that Cubism, which followed, is situated in a 'logical' perspective. Could it then have developed under the brush of artists other than Picasso or Braque? One might raise the same question about Darwin or Einstein. The fact that similar discoveries often happen simultaneously in different places suggests that science holds the key to its own change. Why should things be different in art?

The history of ideas remains to be written; art and science have often evolved similarly, with phases of linear accumulation, stagnation and ruptures marked by ingenious discoveries.

Science in architecture

The Jonas Salk Biological Sciences Research Institute, La Jolla, California, designed by Louis Kahn between 1959 and 1965

Airports, nuclear plants and space stations are some of the twentieth-century architectural creations designed not merely to fulfil their functions, but as emblems of the age. A journalist once wrote about Kahn: 'When architecture reaches the summit of its expression, it arouses a very peculiar thrill, a cool bliss which suddenly invests everything, and through which consciousness of a supremely formulated technical wisdom engages in a conflict with mystical tranquility.'

The unity of art and science

La Dame de Brassempoy, c. 25,000 B.C.
Musée des Antiquités Nationales de Saint-Germain-en-Laye.

The mastery of fire was man's chief prehistoric accomplishment. As the hearth became a place for social gatherings, cooking stimulated interest in the variety of plant and animal life used for paint, poisons and medicines. Fire led to the making of hard clay objects and tools. With the crudest of implements, beautiful sculptures were carved and stunning images produced.

Astronomy, the mother of science, had its roots in prehistoric times, too. It was practised by early hunters and gatherers: notches carved in artefacts suggest that Stone-Age people detected patterns in the motions of the stars.

Astronomy and architecture were linked from very early and would remain so for thousands of years. By definition, architecture is an applied art with functional, technical and aesthetic requirements. It is also a major art in contrast to other applied arts which are generally considered minor.

At the dawn of civilization, people were already conceptualizing in art and science and designing their places of worship to reflect their ideas about the structure of the universe.

The transformation of society during the Stone Age was a major cultural and technical upheaval. Agriculture and the

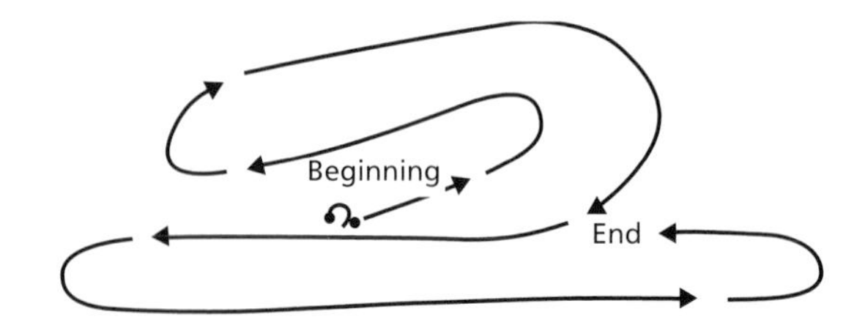

Incised bone record, c. 28,000 B.C.
Interpretation by Alexander Marshack
Long considered as decorations or records of successful hunts, these notches are now believed to constitute a calendar of lunar cycles.
Musée des Antiquités Nationales de Saint-Germain-en-Laye.

way of life that it imposed called for boundary-fixing, which perhaps occurred independently in the Middle East, the Far East and other regions. Stone monuments were the architectural response.

In Europe, thousands of megalithic 'card houses' were scattered across the continent. Long regarded as simple constructions consisting of one colossal flat stone resting on two rough pillars, such structures are now believed to have been the skeletons of richly decorated monuments.

Stonehenge, for example, was built and rebuilt over 1,000 years, with huge and heavy materials that had to be hauled over long distances in ways we still cannot understand. In its oldest known medieval reproduction, the circular complex – poetically supposed to have been masterminded by Merlin – is represented as a flat rectangle with all its triliths complete.

During the Renaissance, scholars thought that Stonehenge was the work of the Romans, for presumably they alone had the required technology.

The monument's compass-like design suggests that it was an open-air observatory. Stones were aligned with celestial events such as the setting sun on the solstices and various phases of the moonrise. Sites like this must have had an important ritual function which historians have interpreted differently, in accordance with the ideas of their own time.

THE TEMPLE OF STONEHENGE, UNITED KINGDOM, BUILT BETWEEN 2750 AND 1500 B.C. (ABOVE). DRAWING OF ITS STRUCTURE ON THE GROUND (LEFT)
This ancient structure is called the 'Neolithic computer' because the fundamentals of astronomy are incorporated into its architectural design.
Photo UNESCO/F. Dunonau

BABYLONIAN MAP OF FIELDS AND CANALS, NIPPUR, C. 1500 B.C.
Geometry was mastered by early Middle-Eastern civilizations. Units of weight and length were legally fixed. Maps were made for tax purposes; this one resembles modern abstract art.
Object B 13885 neg# S4-139370.
University of Pennsylvania Museum, Philadelphia

In the Middle East, more than 3000 years B.C., urban development was well under way. Large settlements called for specialized labour to produce goods and services. Luxury materials used for decoration – marble, flint and alabaster – were actively traded. In one of the largest cities commerce was based on obsidian, a hard volcanic 'glassy' material used for cutting tools.

Sophisticated building techniques were in use. Hand-moulded mud-brick and mortar, sun-dried walls and floors, would soon be covered with coloured plaster and water-resistant tile-like materials.

Although rarely used, columnar pillars were known. Mesopotamian ziggurats, ladder-type buildings – several rectangular stories painted in different colours – were so monumental that archaeologists who discovered them in the nineteenth century mistook them for industrial complexes.

The construction of the Mesopotamian city-states was carried out by a practical, well-organized society in which writing, calculation, sophisticated medicine and astronomy were commonly practised. Although the fragile soil of Mesopotamia caused constructions eventually to collapse, many architectural and other inventions were passed on to the Egyptians.

LIVER MADE OF CLAY INSCRIBED WITH BABYLONIAN CHARACTERS, NINETEENTH OR EIGHTEENTH CENTURY B.C.
This sculpted liver is inscribed with signs, which, when correlated to the movements of the stars, were interpreted by priests for divinatory purposes. Astrology and medicine were also closely linked in Etruscan and Chinese cultures.
© The British Museum, London

Along the Nile, geometry and planning were used for standardizing buildings and for dividing fields. As the Nile, the major highway, was the source of life, water management was mastered at an early stage. The step pyramid at Saqqara (*c.* 2650 B.C.), the world's oldest stone structure, was probably designed on the basis of canal-building experience. Its central monument, which was changed several times, was a technological and an artistic marvel.

The complex of stone and rubble was faced with limestone slabs and richly decorated. Blocks were moulded in imitation of natural materials such as wood pillars and bundles of reeds, suggesting sophisticated earlier wooden constructions. Saqqara's designer, Imhotep, was the first architect ever to have his name recorded and the first in an impressive series of builder-scientists. A celebrated astronomer and healer, Imhotep was deified during his lifetime as the god of learning and medicine.

Soon after, the great pyramids were built at Giza with astonishing precision. Their bases form almost perfect squares with the greatest deviation from a right angle being only half a per cent. The orientation of the sides is exactly north-south and east-west. They were completed within a few decades by methods still partly unknown to us. The core probably rose together with the ramp, yet wheeled vehicles were not used to transport materials. Was granite quarried first, or later on the spot? How were the stones cut to fit with jeweller's precision?

The pyramids had an outer casing of stone, and their tops were covered with a layer of reflective material. At sunrise they were illuminated before anything else around. To the Ancient Egyptians, they may have looked like solar energy stations, similar in concept to contemporary cybernetic sculptures.

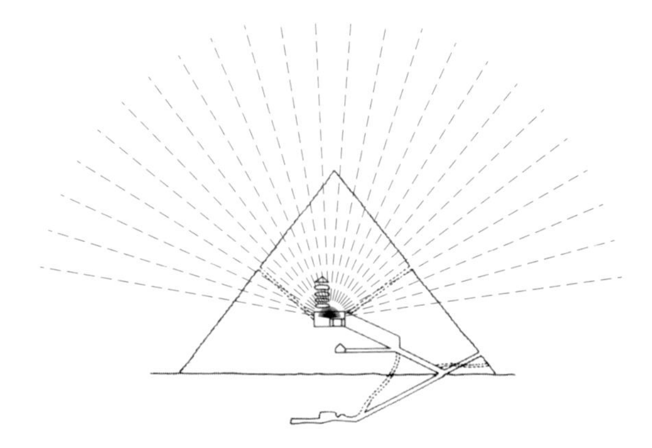

COSMIC RAYS REVEALING THE INTERIOR OF THE CHEOPS PYRAMID

Cosmic rays from space were used by physicists and archaeologists to understand the design and content of the Pharaoh's presumed burial chamber – and the false passages perhaps intended to discourage grave robbers.

COMPUTER MODEL OF THE SPHINX, MARK LEHNER

Originally brightly coloured, the Sphinx's cheeks bore traces of ancient red paint until recently. This computer-generated model is one of many attempts to reconstruct the monument as it looked when it was first built. It is based on close study of a life-size statue of Pharaoh Chephren, who possibly ordered the construction of the Sphinx.

To the Ancient Greeks, pyramids appeared like symbols of geometrical beauty. Later, the Arab traveller Muhammad Ibn Batuta (1304–77) said: '[God Thoth], having ascertained from the appearance of the stars that the deluge would take place, built the pyramids to contain books of science and other matters worth preserving from oblivion and ruin.'

The interpretation of works of art in scientific terms is not peculiar to our time. Links between astronomy and the construction of temples were found through the ages in distant civilizations. Watching the sky, extrapolating cosmic observations to buildings and in some cultures to the human body, was the expression of a broad conceptual system.

CHRYSIPPE, DRAWING OF AN ANTIQUE SCULPTURE, ÉDOUARD MANET, C. 1862

Egyptian monuments contain traces of the grid that guided the sculptor at work. Their canon consisted of precise formal prescriptions that remained unchanged for some 2,200 years. Medieval art had its own canons (this word comes from the craftsman's cane). The technique of plotting drawings on a grid is still in use.

Musée du Louvre D.A.G. © Photo RMN

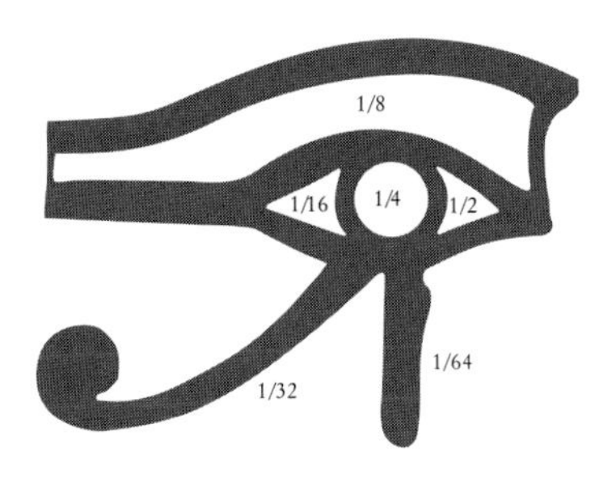

The eye of Horus
The first sequences of the mathematical progression 1/2, 1/4, 1/8, 1/16, 1/32, 1/64 are each represented by a hieroglyph – the combination of which represents the eye of Horus, the god with a falcon face.

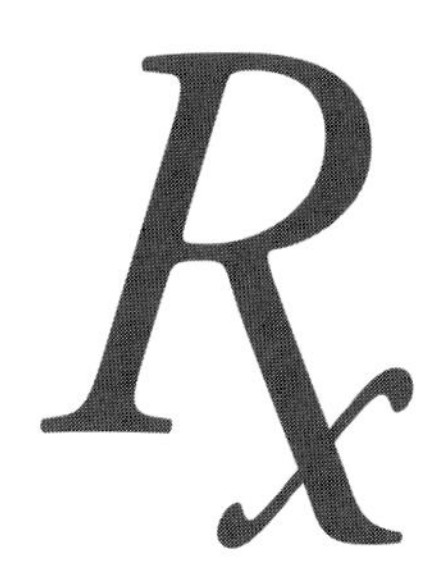

Modern medical prescription symbol
This symbol, still used by physicians to signify a *recipe,* strangely resembles the eye of Horus. Ancient Egyptians knew a great deal about pigments, cosmetics and embalming. They were knowledgeable about diet, massage, hypnosis and contraceptives, and had a pharmacopoeia of hundreds of drugs. In the West, mummy powder would long be considered as a cure for many ills.

Egyptians, whom we tend to admire mainly for the magnificence of their tombs, also made valuable contributions to medicine and astronomy. Like the Babylonians, they used a calendar year of 360 days (and later added 5 1/4 days) and divided it into 12 months, corresponding to the zodiac signs.

Egyptians systematically incorporated that science into their architecture. For example, 160 feet inside the Abu Simbel temple, the statue of Ramses II (*c.* 1304–1237 B.C.) was bathed by sunlight twice yearly on precise calendar days.

Extended knowledge of astronomy prevailed across the entire region. The Talmud relates that, in the Temple of Solomon in Jerusalem, the rays of the sun during the equinox lit the altar by passing through a metal disk in the door. Egyptian monumental relief sculpture, whose design was strictly codified, enhanced the architectural beauty, while assuming its part in the ritual.

Besides their aesthetic appeal, obelisks served as sundials. The most famous of them, Cleopatra's Needle, was used to calculate the time, seasons and solstices.

Observation and reasoning

Across the Mediterranean Basin, an amalgam of ideas from Mesopotamia and Egypt were transcribed using the Semitic alphabet. Commerce with regions as remote as the Baltic resulted in an unprecedented cross-cultural fertilization. Brightly painted Egyptian buildings must have influenced the surrounding architecture in many ways.

The Cretans (*c.* 2500–1100 B.C.) were energized by the arrival of metalworking, pottery and the textile industry. They inherited their neighbours' traditions, but interpreted them in their own way.

Egyptian tomb decorations were thought by the Greeks to be painted inventories of objects for daily use. Homer even described the Egyptians as 'a race of druggists'. Whereas the latter had concentrated on death, the Greeks used science to serve health, developing the art of living to a high degree. They had houses with several rooms and tiled roofs, running water and luxurious bathrooms.

Despite their condescension towards the Egyptians, the Greeks are believed to have adopted Imhotep as their god of medicine. They gave him the name Asklepius, and his daughter-wife Hygeia was the goddess of health. The practice of medicine extended by Hippocrates (*c.* 460–377 B.C.) became an art based on technical recipes.

Before the Ancient Greeks, science was a loose collection of observations used for practical applications. The Greeks, who developed a keen awareness of space – probably with their navigational skills – were responsible for the birth of science, and even of science for the sake of science.

Thales of Miletus, regarded as the founder of natural philosophy (*c.* 625–547 B.C.), studied astronomy in Mesopotamia and stunned all who knew about it by correctly predicting a solar eclipse. In Egypt, he learned land surveying from which he deduced geometry.

According to Thales, the universe was made of a physical substance, water. From his time on, philosophers sought to understand the basic mechanisms of nature by the use of analogy and reasoning. They introduced a systematic approach in all avenues of creation.

The great philosophical awakening was accompanied by a formidable architectural movement. The Egyptians used the

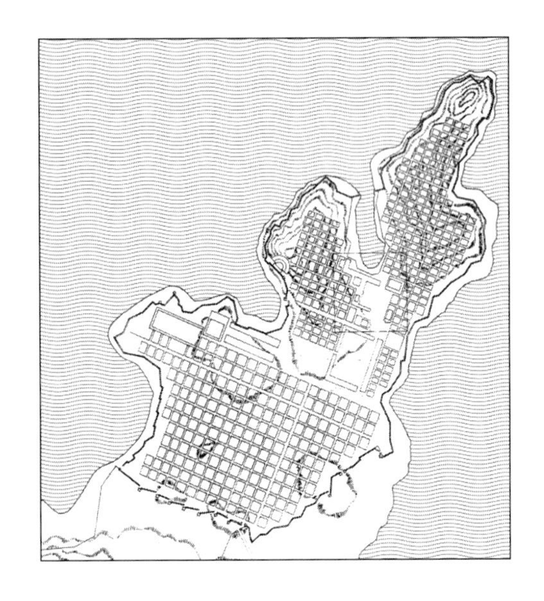

PLAN OF THE ANTIQUE CITY OF MILETUS
The Greeks learned mathematics from the Mesopotamians and the Egyptians, and used it in their urban and architectural plans. A map of Manhattan would not look very different.

grid as an aid in architectural design, but the Greeks extended orthogonal planning to the layout of entire cities. Tapered columns, angle contraction and other design tricks used for visual effectiveness were current in the Middle East; the Greeks adapted them to a system of 'ideal' proportions based on the Golden Section. This became fundamental to architecture in a unique way.

The union of body and spirit resulting in architectural forms related to human anatomy was typical of the Ancient Greeks. Builders were capable of transforming mathematical concepts into architectural delights to please the senses. They created art for the sake of art, so to speak.

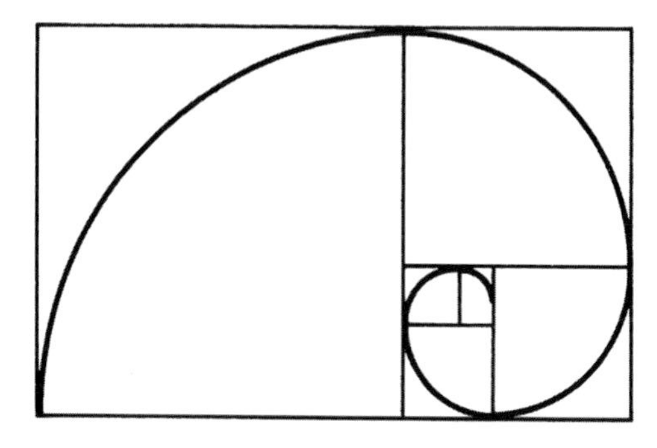

THE GOLDEN SECTION
The Greeks were fascinated by the Golden Section – a line or a rectangle divided into segments in such a way that the smaller one has the same ratio to the largest, as the largest has to the whole. They were not the first and would not be the last to exploit its beauty. The Golden Section is found in the Cheops pyramid and Chartres Cathedral, as well as in plant growth patterns.

THE THREE GREEK ARCHITECTURAL ORDERS
The Greeks used three orders as modules: Doric, based on early Eastern models, Ionic, and later Corinthian, which defined the proportions of the whole building. Concerning the strict rules of the Doric order the Roman architect Vitruvius said: 'Of whatever thickness they made the base of the shaft, they raised it along with the capital to six times as much in its height. So the column began to furnish the proportions of a man's body strength, and grace.'

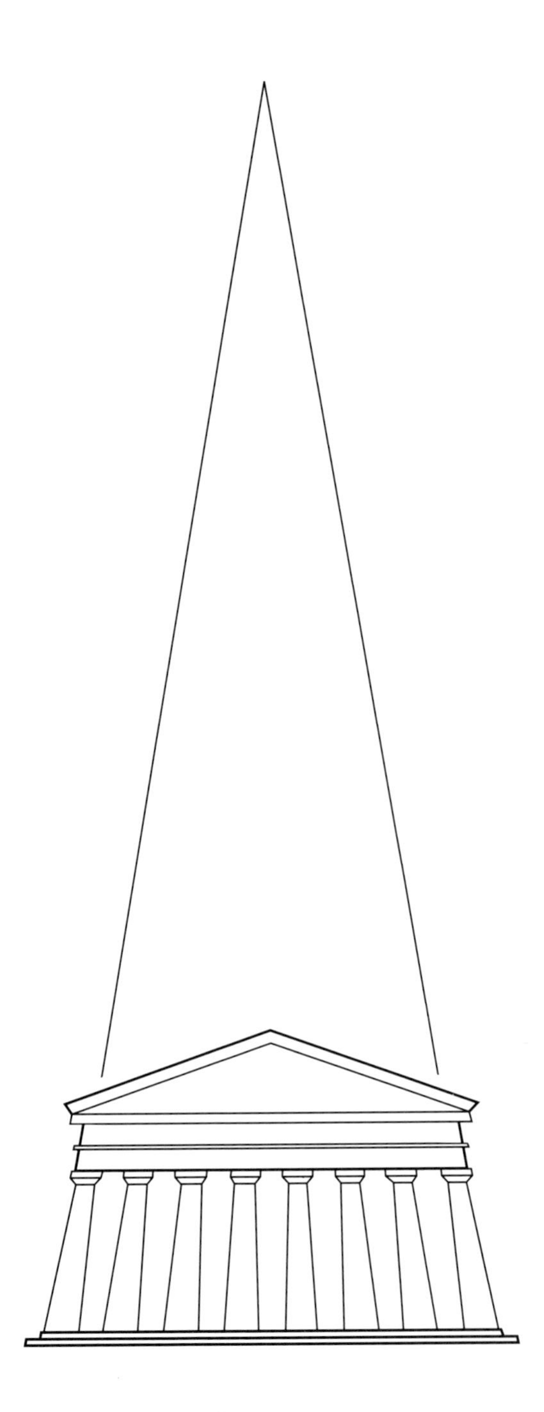

CONVERGENCE OF THE PARTHENON COLUMNS,
PHILIPPE COMAR

The Parthenon's forty-six columns theoretically converge at a point 2,000 m high (which would be situated at infinity if the columns were parallel). The Greek builders were aware of the illusion which makes a bright object appear to be bigger than a dark one. To offset this effect, columns viewed against dark walls are thinner than the corner columns, which are seen against the sky.

La Perspective en Jeu, *Découvertes Gallimard, 1992*

Early Greek temples were like shrines containing a statue. The cult of human-looking gods institutionalized by the Olympic Games (ninth century B.C.) spawned a new form of architecture. The typical Greek construction, supported by stylish columns, became a place of public worship. Its wide porches and brightly painted friezes encircling the building beckoned from all sides.

Stone blocks were shaped with exacting precision and fitted together without mortar. No classical Greek architectural plan has been found, but it is evident that temple design based on ideal proportions was standardized. Only small variations were accepted in the overall scheme.

Ideal proportions, rather than the simple grid as used in Egypt, played a key role in determining the appearance of sculptures and buildings in Ancient Greece, as well as shaping theories. Plato (427–347 B.C.), who was interested in the intrinsic beauty of forms rather than in fashionable design, believed that geometry provided the key to the mysteries of nature, science and art alike.

Making the world intelligible in mathematical terms was one of the great advances of human thought, bringing abstraction into play. Order, hierarchy, ethics and aesthetics were intricately interwoven. Plato, who had a special interest in education, advocated: 'Let your children's lessons take the form of a game. Learning through play is linked with sympathy, and conformity with beauty and reason.'

Aristotle (384–322 B.C.) thought that facts should prevail over elegant concepts. His pupil, Alexander the Great (356–323 B.C.), took geographers and engineers along on his military campaigns and sent back plant and animal specimens from wherever he travelled. (Legend has it that the great empire-builder was an amateur naturalist and had a special glass vessel built to facilitate his underwater observations.)

Alexander's collection – in which every creature was believed to have a function in the Great Design – prompted Aristotle to compile an encyclopedia that was to contain all knowledge. Thereby he stimulated the switch from speculative to empirical thinking. In addition, he provided a framework for the discussion of philosophy and disciplines such as logic and physics. One of his many noteworthy achievements was to prove formally that the earth is round.

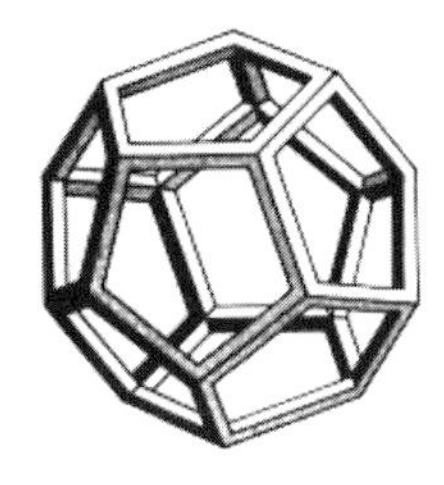

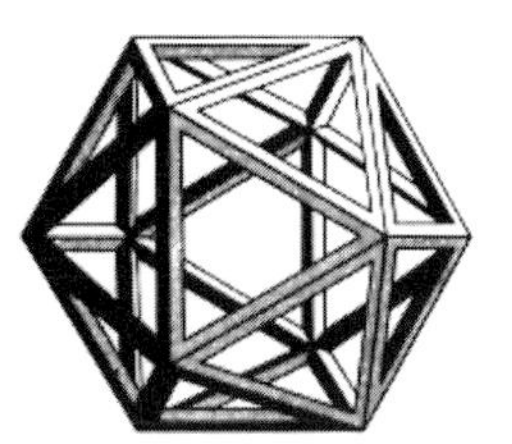

THE PLATONIC BODIES
A theory of solids was based on aesthetic considerations. Philosophers enamoured with the concept of symmetry supported the argument that there had to be a continent on the other side of the earth, to maintain its equilibrium. Almost 2,000 years later, James Cook, while searching for it, found Australia!

Some even precisely calculated its circumference (just 15 per cent greater than the actual measurement), and others went so far as to oppose Aristotle's geocentric view by asserting that the sun is the centre of what we now know as the solar system. Unfortunately this theory was quickly forgotten.

Greek scientists created and followed trends, just as artists do today. Some believed that the four elements (earth, water, fire, air), thought to constitute the universe, could be altered by forces such as love or strife. The elements, themselves a subdivision of the general cosmogony, were viewed as the constituents of the human body.

Natural philosophers made analogies to transformative processes in crafts, such as pottery and metallurgy, to explain the functioning of the body and the formation of the earth. The fundamentals of Western alchemy were established by these observations.

A vast amount of information was generated and learning eventually entailed organized study in schools. New architectural concepts emerged: Stoas, buildings with colonnades where students could talk and walk, lycea and academies. The Museum of Alexandria was a teaching centre that attracted scholars of all types, and employed about a hundred state-paid professors.

Empirical methods in science found echoes in art's new freedom and naturalism. Realistic sculptures decorated multistorey buildings. At the same time, pragmatic inventors, forerunners of today's engineers, Archimedes of Sicily (*c.* 287–212 B.C.), Philon of Byzantium (*c.* third century B.C.) and Heron of Alexandria (first century A.D.), invented mechanical devices used for stage scenery as well as for military purposes.

PORCH OF THE MAIDENS, ERECHTHEUM, THE ACROPOLIS, 421–405 B.C.
Sculpture was an integral part of this structure, which related to human proportions. Tinted wax was used to colour the hair, lips and costumes of the figures.
Photo UNESCO/D. Roger

Improved weapons with high-velocity projectiles also influenced architectural design.

Meanwhile, in Rome, networks of paved roads were being built. Across the expanding empire a large-scale uniform construction programme was carried out – as Rome had a monopoly over naturally occurring materials such as marble and travertine.

The Romans exploited mines in England, imported silk and spices from the East and cereals from Russia. In the south of France, mills produced enough flour to meet the year-round needs of almost 100,000 people; most of it was exported to feed the troops.

The Romans, whose strengths were essentially administrative and organizational, made two major contributions to architecture: they developed the potential of concrete – a light-weight, fire-resistant material composed of rubble, water and mud, produced on building sites; and they extended the use of the arch which enabled them to create the biggest interior spaces made until that time.

The Ancient Greeks used arches for small scales and found them unappealing. Under the Romans, arches became ubiquitous, being used in new classes of buildings and in improved versions of traditional ones. Basilicas for gatherings, hospitals for soldiers, amphitheatres and hippodromes, and public baths decorated like contemporary art galleries, sprang up. Mega-projects such as dams, aqueducts, canals and tunnels, some of which are still in use, were built throughout the empire.

In Rome, the Colosseum (*c.* first century A.D.) accommodating more than 50,000 people, remained the world's largest amphitheatre until 1914 (the year of the construction of the Yale Bowl). Similarly, the capital's crowning monument, the Pantheon (*c.* 30–126 A.D.) remained the largest dome for centuries. Nine-tenths of it was built of concrete, a material that had been upgraded and refined over a period of two hundred years. Over its cement core, the Pantheon (the symbolic home of all the gods, as its name indicates) was faced with luxurious materials. Egyptian porphyry, granite and the finest Greek marble covered the massive building.

Vitruvius (first century A.D.), a Roman engineer who admired the Ancient Greeks but failed to appreciate the building talents of his peers, is the author of a text that was to become the construction bible of the Renaissance. He gave the first account of architectural acoustics and explained that sound is caused by the vibration of air. Vitruvius, who also theorized on astronomy and waterwheels, wrote: The architect 'must be educated, skilful with the pencil, instructed in geometry, know much history, have followed the philosophers with attention, understand music, have knowledge of medicine, know the opinion of the jurist, and be acquainted with astronomy and the theory of the heavens.' A tall order!

NAVAL GAMES, SERGE STROSBERG, 1998
The Romans built an aqueduct to supply a large artificial lake for mock naval battles. They flooded amphitheatres such as the Colosseum. Its circular, richly decorated structure – each storey being designed according to one of the Greek orders – admitted and disgorged audiences through miles of interior stairways leading to seats around the arena.

Roman scientists also took their cues from the Greeks. The astronomer Ptolemy (second century A.D.) drew maps of various countries presented in his *Geographica* and charted over 1,000 stars in his *Almagest* (*Al Majisti* as the Arabs respectfully named it). Their given positions would go unquestioned until late in the Renaissance.

Galen (second century A.D.), a most famous physician, was particularly interested in Greek art and made reference to *The Canon* written by a sculptor in these terms: 'Beauty arises not in the commensurability of the elements, but in that of the parts, such as the finger to the finger, and of all the fingers to the palm and the wrist, and of these to the forearm, and of the forearm to the upper arm, and in fact from everything to everything else.'

From his dissections of monkeys and pigs, Galen extrapolated information on human anatomy that remained unchallenged until findings based on the dissection of human corpses were published over 1,000 years later. The Romans did a fine job of disseminating knowledge passed down from the Ancient Greeks, but they adopted their content without the reasoning methods. Despite their many achievements, the Romans are remembered as craftsmen and experts in technology. Their culture never attained the brilliance of the Ancient Greeks and Italians of the Renaissance who cultivated art and science with equal zeal.

COPY OF A PAGE OF A BOTANY TREATISE, TENTH CENTURY

Discorides, a physician of Greek origin who served in Nero's army, described several hundred plants and their medicinal virtues. His work would be abundantly commented on by Arab scholars.

Österreichische NationalBibliothek, Vienna.
Photo Bildarchiv, ÖNB

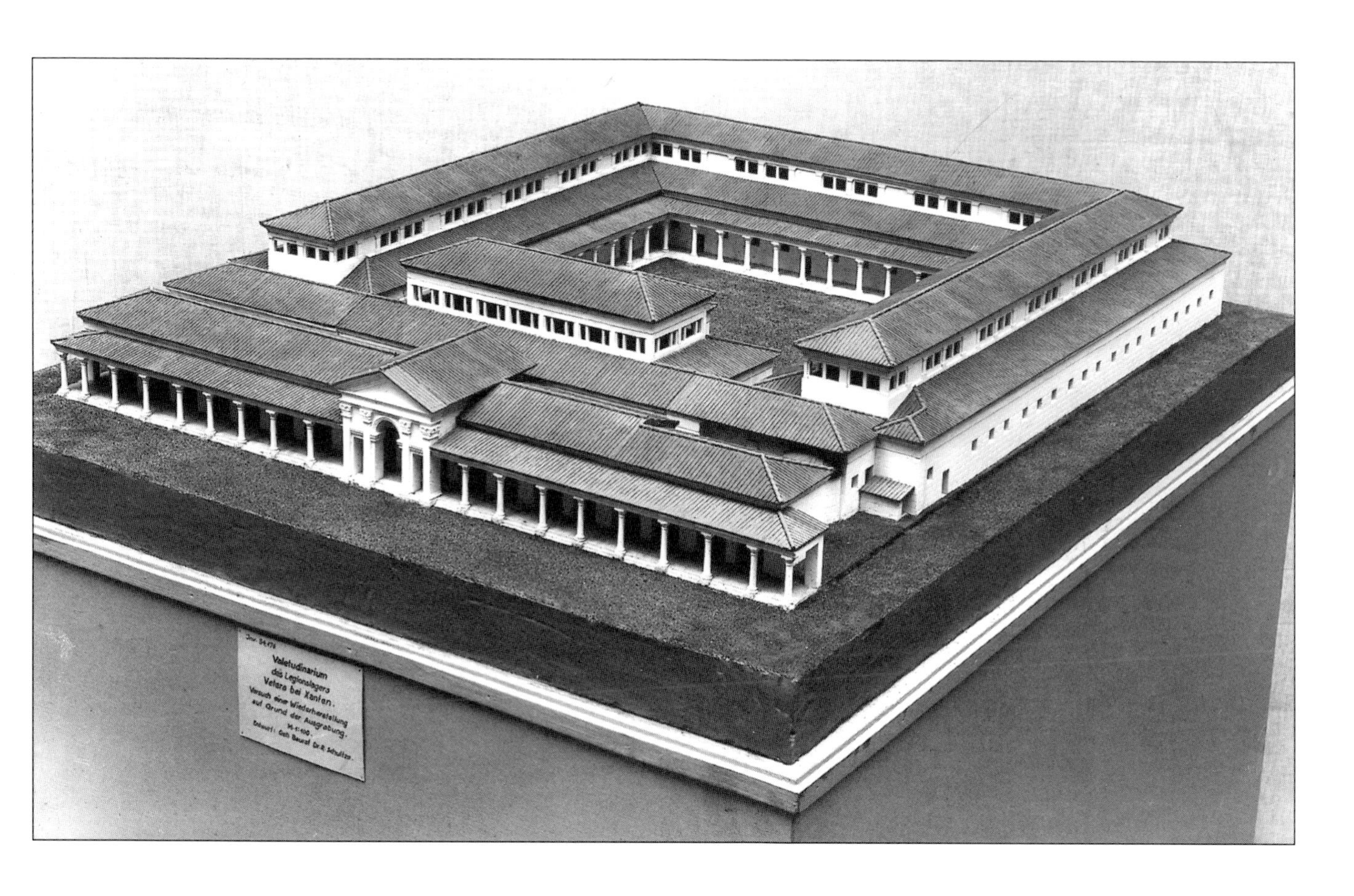

Model of a hospital for Roman troops

Rheinisches Landesmuseum, Bonn

Faith transcending building logic

As the power of Rome declined, the pursuit of most forms of knowledge ground to a halt. Exceptionally, architecture made headway as the rituals of the Christian Church required large gathering spaces. The emphasis was on the interior space, spiritually and architecturally. In Constantinople, which became the Christian capital, Hagia Sophia (sixth century) rose even higher than the Pantheon in Rome, yet it seemed weightless inside. Its construction was a brilliant feat, since the techniques of building with concrete had been lost.

An immense celestial sphere was tiled with scintillating mosaics which astounded visitors: 'Without external sunshine . . . the rays of the sun emerged from inside.' The challenge of placing this dome over the world's largest square building was entrusted to Anthemius of Tralles, a well-known author of treatises on geometry. Like the scientist-astronomer, Imhotep, Anthemius created a watershed in the history of architecture.

Interior of the Pantheon in Rome,
Giovanni Paolo Pannini, c. 1734

The Pantheon in Rome, which served as an astronomical observatory, became the primary model for Christian and Islamic architecture. The construction was made by use of stacked concrete compartments, which permitted drying of huge volumes of material needed to sustain the ceiling; lighter material was used near the circular opening at the top. In the Paris Panthéon, built in the eighteenth century on the Roman model, Léon Foucault gave the first material demonstration of the rotation of the earth.

Oil on canvas. Samuel H. Kress Collection.

1.283 .991 (501/2 39). Photo Richard Carafelli

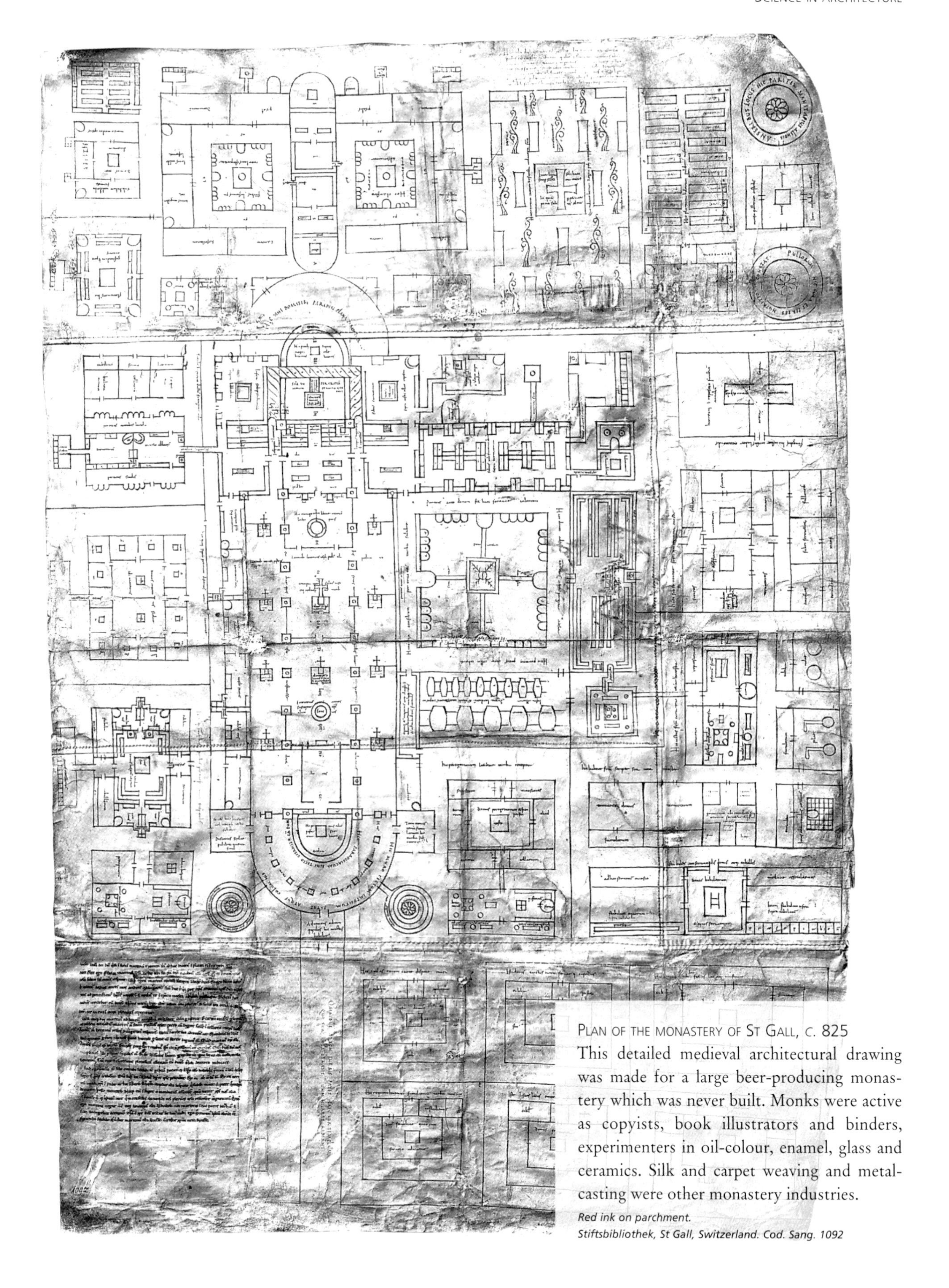

PLAN OF THE MONASTERY OF ST GALL, C. 825
This detailed medieval architectural drawing was made for a large beer-producing monastery which was never built. Monks were active as copyists, book illustrators and binders, experimenters in oil-colour, enamel, glass and ceramics. Silk and carpet weaving and metal-casting were other monastery industries.

Red ink on parchment.
Stiftsbibliothek, St Gall, Switzerland. Cod. Sang. 1092

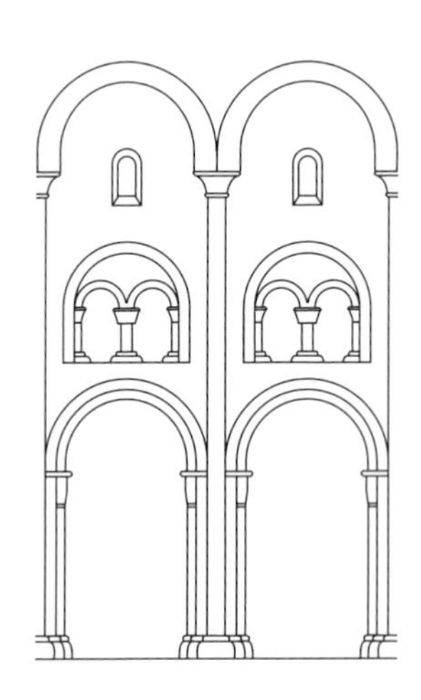

Romanesque style

Comparison of Romanesque and Gothic (opposite page) nave elevation, Mikiko Noguchi
The pointed arch, vaulted nave and flying buttress enabled architects to realize their boldest visions. It is said that between 1170 and 1270, a quarter of the gross national product was spent in labour and material for some eighty Gothic cathedrals in France.

During the ninth century, Charlemagne's taste for Roman splendour led to the development of the Romanesque style. Across his western empire, he blended the refinement of the Byzantine church with the architectural logic of Antiquity into a hybrid form combining Roman arches and Greek columns.

Characterized by impressive masonry, Romanesque constructions grew up along pilgrimage roads between the ninth and twelfth centuries. Monasteries ran prosperous craft workshops and were successful in livestock-breeding, farming and forestry. Technical progress resulted in improved productivity. The horse collar made traction easier (placing the strain on the shoulders instead of the neck), and a more sophisticated plough lightened the farmer's burden. During this agricultural transformation, 5,000 watermills were listed in the *Domesday Book* – an unprecedented large-scale census conducted in 1086 in England, at the request of William the Conqueror.

In France, the Cistercians were one of the most technologically advanced religious communities. At Cluny, the Benedictine order was the cultural as well as the economic leader. At the height of its power, several hundred monasteries across Europe belonged to Cluny's network.

One of Cluny's anonymous architects believed that the visual arts revealed a link between music and the harmony of the universe. Liturgical music developed in concert with architecture. Art and science being progressively rediscovered and secretely commented on, monastic building developed in parallel.

Characterized by multiple arches and crowded polychromic sculpture, architecture often followed geometrical patterns. Audacious experimentation was taking place. In medieval buildings, the weight of the stone vault rested directly on the supporting lateral walls, limiting ceiling height and wall openings. Attempts were made to build churches with higher and more expansive ceilings, but walls and arches were rarely straight, and the structures often collapsed. New construction methods were desirable.

The great Gothic awakening started with the use of the ribbed vault over a multistorey nave. This type of vault, with small triangular sections to fill, was easier to build than a unitary mass. What is more, the pointed arch could display angle variation as round arches could not. The flying buttress became yet another tool in the conquest of verticality, providing lateral support. Possibilities arose for the creation of huge interiors reaching new heights by dispersing weight.

The powerful abbot, Suger (1080–1151), directed the construction of St Denis, where the kings of France are buried. He was guided by his belief in harmonious proportions, and by a personal interest in the symbolism of light. Suger fully integrated various techniques which he observed as he travelled; he was also, incidentally, a collector of gemstones and coloured glass.

His greatest feat was in synthesizing all the diverse visual and structural elements of what came to be known, centuries later, as the Gothic style, into a unified and spiritually expressive whole that was structurally sound. Word of Suger's achievement spread and, within a hundred years, examples of the Gothic style could be found all over Europe.

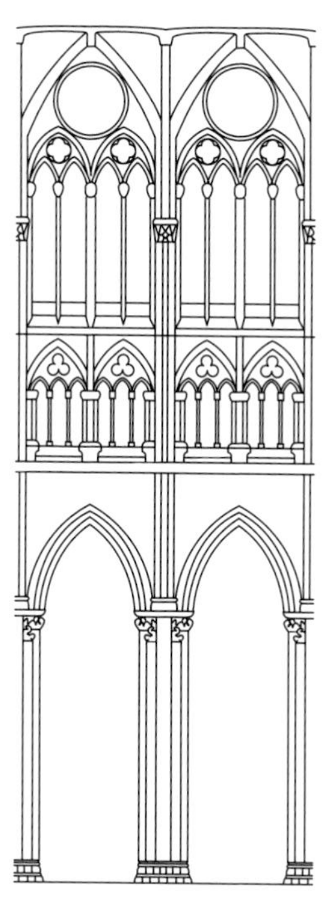

Gothic style

During the nineteenth century, the French architect Eugène Viollet-le-Duc, advocating the neo-Gothic style, described Gothic as an engineer's art that drew inspiration from practical utility: 'Nature has not found the unfindable, the absurd . . . it proceeds as we would, adjusting bodies to functions . . . an example we should follow, when we pretend to create using our intelligence.'

Nowadays, scholars are no longer convinced of the rationale of Gothic structures. The quest to turn churches into containers of divine light and space suggests that art had gained supremacy.

The same techniques accelerated the development of secular Gothic art in a rapidly expanding Europe. Excited by innovation, a wealthy middle class financed research in alchemy, mining and metal-working. Flourishing trade accelerated urbanization, which gave rise to new forms of architecture, such as storage depots, shops and town halls.

New standards of comfort were reflected in interior decoration: embroidered cloth, wood panelling and carving. One of the most celebrated examples is the Bayeux Tapestry, celebrating the Norman victory over England (in which the comet is the same as the one described 500 years later by Edmond Halley, the English astronomer: its passage would confirm Newton's theory of universal gravitation).

Both art and science were ostentatiously sponsored by patrons who wanted to be remembered. Thus, symbols of power and high technology, such as mechanical clocks, were incorporated in the exteriors of major cathedrals. Time, until then conceived as

COMPUTER-GENERATED IMAGE OF THE ABBEY OF CLUNY, 1993
The lavishly decorated abbey of Cluny was destroyed following the French Revolution in 1789, and Napoleon's edict in 1810. Three-dimensional computer images have been generated from the work of the architect Conant, who spent decades deciphering the plans.
© IBM-Artway

cyclical and determined by the positions of the sun and the moon, became a continuum, tracked by a scientific device.

This innovation heralded a profound spiritual change, gradually replacing a qualitative perception of the world with a quantitative one. But, before this took place, Europe lost about one-third of its population to the plague, leaving wealthy survivors aspiring to build a different future.

GOD THE GEOMETER, THIRTEENTH CENTURY
Medieval architecture was deeply steeped in antique symbolism. The desired harmony – a perfect relationship of different parts in terms of ratios – was the source of beauty according to which 'the divine reason ordered the universe'.
Österreichische NationalBibliothek, Vienna.
Photo Bildarchiv, ÖNB

THE ASTRONOMICAL CLOCK OF STRASBOURG CATHEDRAL, FIFTEENTH CENTURY
In medieval monasteries, work and prayer were regulated by the ringing of bells; days were divided into periods based on the total time elapsed between sunrise and sunset . . . but this duration varied from day to day. The necessity for telling time became widespread, and common standards were established. The day was divided into twenty-four equal segments – the hours – shared by all. In today's industrialized world, clocks controlled by atomic vibrations are accurate to within a millionth of a second.

Photo © LL-Viollet

Distances and encounters

While the West was about to take a new direction, other societies in various parts of the world lived according to elaborate models. In the Americas, successive civilizations flourished for over 2,000 years (beginning around 1000 B.C.). They conducted immense building projects: road networks tens of thousands of kilometres long crossed the formidable terrain of the Andes mountains.

Masonry blocks weighing 100 tons were used to build monumental structures which functioned as giant platforms for ritual practices. They were carefully aligned with the stars, and in close proximity to the heavens. Pre-Columbian pyramids bear resemblance to their Ancient Egyptian counterparts. There are similarities in the understanding of mathematics and astronomy attained by these distant cultures. Incidentally, the dry air of the Andes, like that of Egypt, also favoured rites of mummification.

In the Americas, fundamental technologies such as wheeled vehicles and the use of iron were unknown; so were horses as beasts of burden. When *conquistadores* invaded the continent on horseback carrying gunpowder and metal weapons – and illnesses to which the local people had no immunity – the result was near extinction.

One must admire the Americas' early inhabitants who conceived a universe 100,000 years old. The Mayas even concluded that time had no beginning – a strikingly abstract notion. Western scientists, including Newtonian thinkers, believed until well into the eighteenth century that the world had been created in 4004 B.C., during the great flood described in the Bible.

INDIAN STATUETTE, SOUTH AMERICA
Deformed faces testify to the interest shown by American Indians in medicine, a long time ago.
Staatliche Museen zu Berlin. Museum für Völkerkunde. © bpk

NAZCA LINES DEPICTING A WHALE, PERU, C. 500 A.D.
Dark stones were removed over tens of kilometres, uncovering the lighter colour of the earth. These drawings representing animals or geometric shapes – recognizable only from the air – must have required measuring skills. Maria Reiche, fascinated by mathematics, studied the astronomical references apparently contained in these figures which were possibly used to predict the outcome of agriculture. Pre-Columbian Indians grew numerous species of plants. More than half of the food we eat is derived from their products and methods.

Mask, Ivory Coast
Some masks show faces with signs of leprosy, mycosis or parasitosis. African statuary greatly inspired Western artists.
With permission of Me Clouzot. Photo Patrick Rimond

It is now common knowledge that societies in Africa and the South Pacific had art schools with masters, and an art story of their own. Yet, although their works are widely appreciated in the West for their aesthetic power, the meaning of these objects is shrouded in mystery.

Specialists are trying to piece the puzzle together, tracing stylistic affinities and material components back to their sources. But colonialism has wiped out most traditions, rites and ceremonies for which staffs and figurines served as accessories. The same is true of medicinal potions and practices. Interest has been kindled in the traditional science of these lands. One of the direct results is the burgeoning discipline of ethnopharmacology, which is bringing to light potent drugs and antidotes to various poisons.

Ancestral know-how may be of use in a number of other areas, not least agriculture and environmental practices. Ironically, hi-tech computer analysis is now employed to record whatever is left of this body of knowledge.

Fetish from Central Africa, Sankuru-Huana
Figurines representing healers carrying medicine bags, dolls, herbs and utensils reveal the deep link between medicine and artistic representation.
Staatliche Museen zu Berlin. Museum für Völkerkunde.
© bpk. Photo W. Schneider-Schütz

CÉLÈBES, MAX ERNST, 1921

The contribution of African art to modern Western art movements such as Fauvism, Cubism or Surrealism is immense. Picasso considered the link between modern and tribal art, discovered at the turn of the century, to be comparable to that between the Renaissance and Antiquity. Max Ernst seems to have been inspired by a granary shape when he painted this elephant.

Tate Gallery, London. Tate Picture Library

AFRICAN GRANARY

Analysed by R. Penrose. University of Newcastle upon Tyne, 1972. Photo H. H. Schomburgh

Less than one century after the death of Muhammad (570–632), the Arabs conquered a huge swathe of land from central Asia to Spain, passing through North Africa. In this way, they brought about a synthesis of Eastern and Greek culture. The works of authors such as Euclid and Ptolemy were abundantly translated by the Arabs, who also studied Indian mathematics.

In Spain, the Cordoba Mosque and the Alhambra Palace were masterpieces of geometrical harmony that had a strong influence on Christian architects: the reverse had occurred in Constantinople years earlier, when the Hagia Sophia church had become the model for builders of mosques.

Arab palaces were surrounded by magnificent gardens. Plant motifs were used in decoration containing flowers, fruit trees and medicinal herbs – botanical textbooks were plentiful. The Arabs designed precision instruments, glass

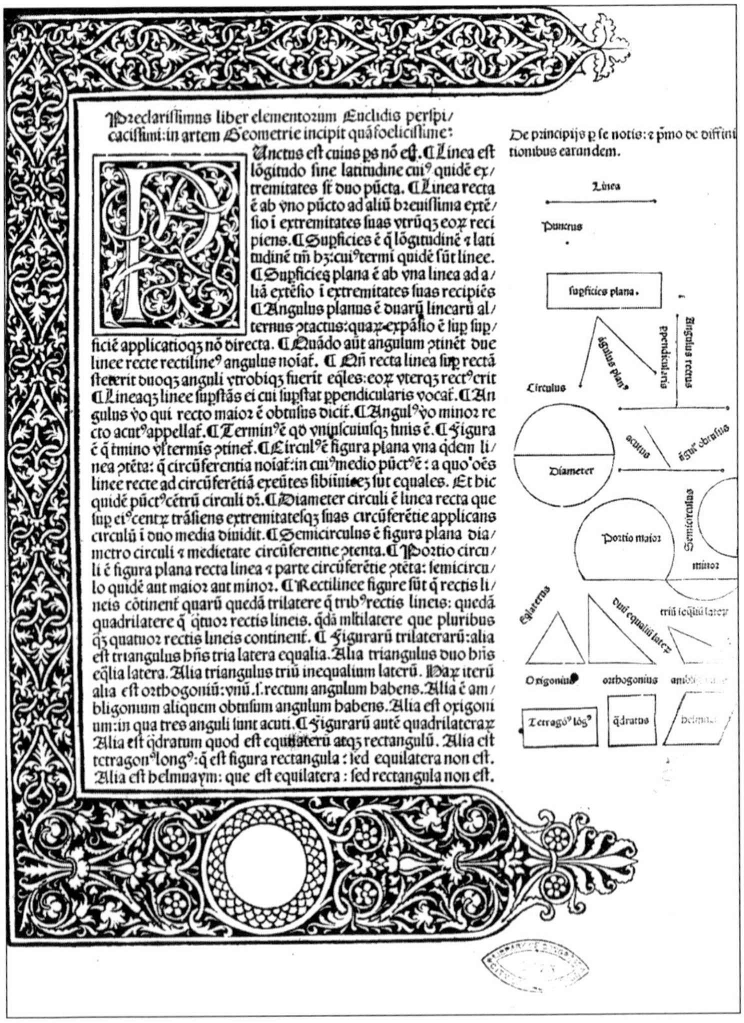
Preclarissimus liber elementorum Euclidis perspicacissimi: in artem Geometrie incipit quã foelicissime:

De principijs p se notis: & pmo de diffinitionibus earandem.

FRAGMENT OF EUCLID'S *ELEMENTS*, LATIN TRANSLATION BASED ON AN ARAB VERSION

European languages are sprinkled with words that have Arabic roots: zero, cipher, hazard, almanac, algebra, alchemy, alcohol, etc. Arabic numerals allowed for written calculations which were impossible with the Roman system. Euclidian geometry was translated by the Arabs and formed the basis of one-point perspective. The *Elements* became almost as well known as the Bible.

E8616 Library of Congress, Washington, D.C.

mirrors and lenses, and excelled in astronomy. They began structured chemical studies as distinct from alchemy; they knew the principles of distillation and experimented with acids and oil paint. Pharmacology was becoming a field in its own right, related to but separate from medicine. (In Baghdad, for example, there were over 1,000 licensed physicians practising around the tenth century.)

PERSIAN MINIATURE, FIFTEENTH CENTURY
These physicians are shown holding a clinical consultation; they carry a small book of diagnostic information on a belt around the waist. Taught in Salerno, Montpellier and Padua, medicine was the cornerstone of some of the early and most prestigious universities.
Österreichische NationalBibliothek, Vienna. Photo Bildarchiv, ÖNB

Muslims, Christians and Jews cohabited more or less harmoniously for hundreds of years. When Christians reconquered Spain, the fall of cities such as Toledo (1085) with its huge library, contributed enormously to the cultural revival of the West. It was Ancient Greek knowledge, transmitted by the Arabs, that persuaded philosophers such as the Jew Maimonides (1135–1204) and the Christian Thomas Aquinas (1225–74) to reconcile their faith with Aristotelian logic. This bold and potentially heretical step was the first to allow the separation of the religious and secular domains. Science has established itself ever since as an integral part of cultural values.

MEZQUITA, CORDOBA,
MAURITS CORNELIS ESCHER
Forbidden by their religion to represent the human figure, the Arabs developed an alternative art form based on mathematics. In the Cordoba Mosque, arches are aligned in a captivating manner as in Romanesque architecture. Around the twelfth century, there were more than 100 mosques in Cordoba.

The Arabs probably learned much from the Far East, although how much, and by what means, is difficult to determine. In many domains, the Chinese were centuries ahead of the rest of the world. Agricultural inventions, the mining of coal and the development of materials such as iron, were evident in China long before they became known in Europe.

Huge wooden constructions were built thanks to hard metal tools. The Chinese designed metal bridges and excavated underground wells 1,000 metres deep. They even knew of natural gas and oil as combustible and lighting materials.

Why were some of their inventions never adopted by the Arabs? Why did it take so long to pass them on to Europe? Which ones were reinvented *ex nihilo*? And, in particular, why were they not used more to stimulate development of Chinese society itself?

One answer, among others, is that Asian philosophy emphasized meditation rather than interrogation, which is familiar in Western thinking. Chinese innovations tended to be isolated events rather than a catalyst for a chain reaction. Deeply impregnated by Confucian thinking – which dated back to the sixth century B.C. and commanded respect for ancient values – the Chinese took pride in continuity, an attitude which is reflected both in their science and their art.

Architectural canons were fixed at an early stage and buildings were reconstructed in a similar style every couple

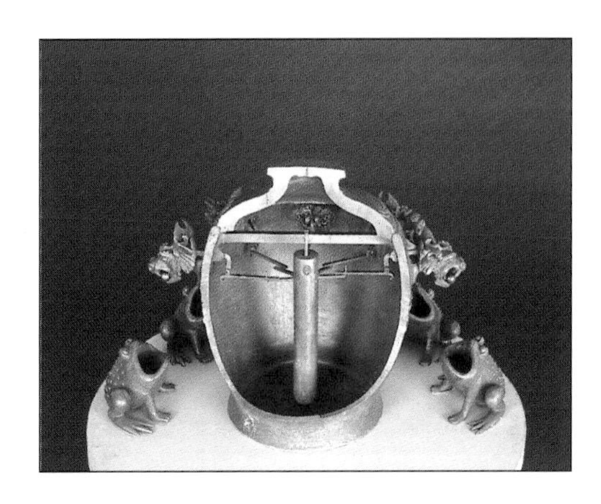

BRONZE SEISMOGRAPH, COPY OF A SECOND-CENTURY CHINESE ORIGINAL

This artistic masterpiece is a scientific instrument. Vibration causes a bronze ball to fall from the dragon's teeth into the toad's mouth beneath. A deep sound is produced at the slightest earth tremor. Even a distant earthquake is registered and the direction of its epicentre is also indicated.

Science Museum, London.
(Source) Science & Society Picture Library

of decades. Thus construction did not evolve substantially either. Nevertheless, across Asia independent styles flourished.

Art of all kinds was formally codified, leaving little scope for subjective expression. Still, within its tight framework, it announced itself subtly and – judging from the impulsive nature of that untidy concept called 'creativity' – probably ineluctably.

Far-Eastern inventions rarely found widespread applications as they were created for the rulers' amusement. According to the Italian writer, Italo Calvino, the Mongolian emperor Kublai Khan (1215–94), who founded the capital of Beijing, said: 'I have neither desires nor fears and my dreams are composed either by mind or by chance.'

Marco Polo (1254–1324) would have replied: 'Cities also believe they are the work of the mind or of chance, but neither the one nor the other suffices to hold up their walls. You take delight not in a city's Seven Wonders, but in the answer it gives to a question of yours.'

Meanwhile, curiosity about the material world and its place in the cosmos impelled the West to march at a faster speed. The persistent questions of merchants, discoverers, inventors, scientists and artists brought about the Renaissance.

IRRIGATION MAP OF THE RUINS AT ANGKOR, CAMBODIA, MIKIKO NOGUCHI

This gigantic Khmer complex built between the ninth and thirteenth centuries comprises many temples – a number of which, buried in the jungle, were found by means of satellite. Little survives of the residential splendour, but remaining reservoirs and temples indicate that it must have been one of the largest cities of the ancient world. Angkor Wat, the main building, is aligned with the sunrise at the spring equinox and its dimensions correspond to astronomical cycles. Containing emblems of fertility, it was conceived as a microcosm of the universe.

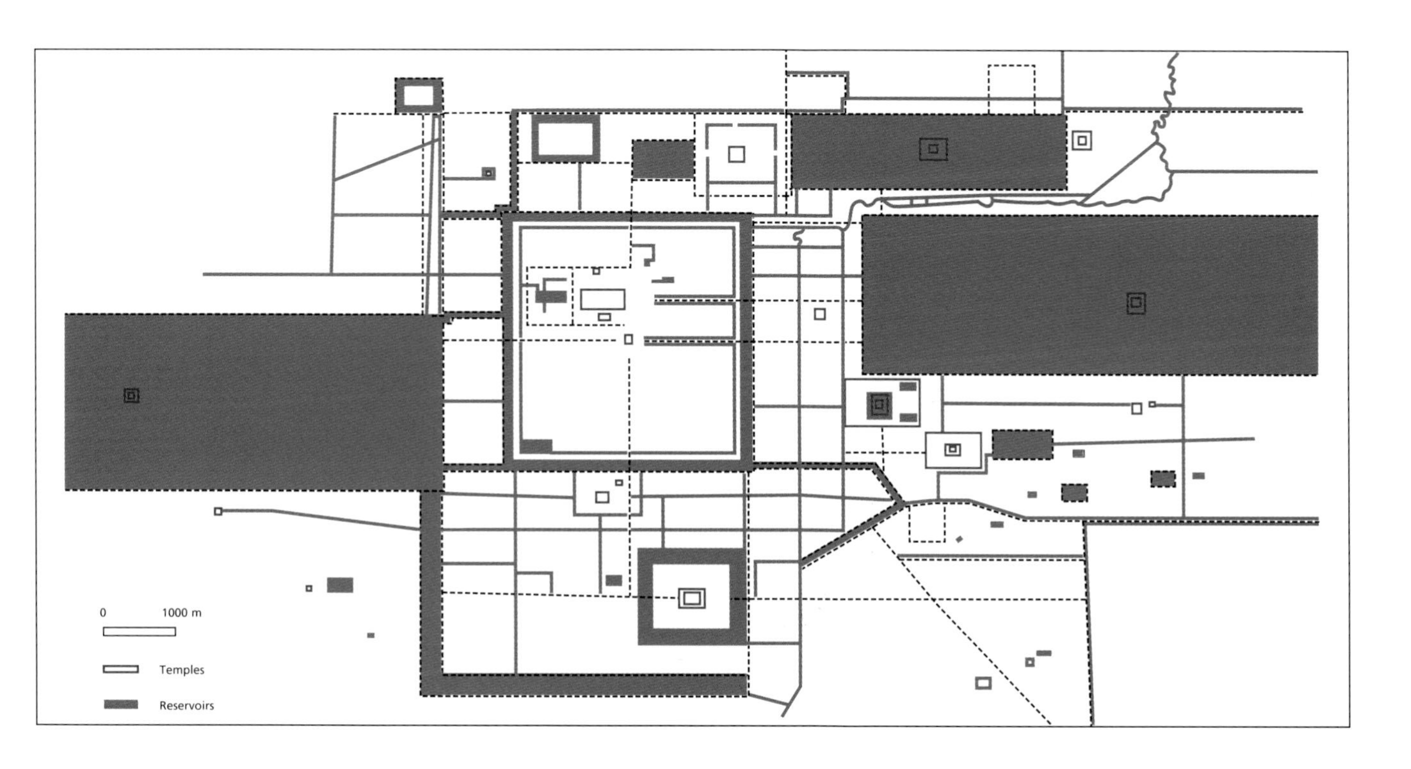

In ancient times, knowledge was acquired directly through perception. In the West, the earth was seen as the centre of all things, with the sun and the stars, the planets and the moon circling around it in fixed orbits. The common scheme envisaged was spherical, and thought to have been set in motion by a primary force.

Over time, the number of heavenly bodies became a subject of debate, and in the Middle Ages it was concluded that the earth was flat. In general, the ancient earth-centred cosmogony endured until the Renaissance. The earth itself was pictured as three continents surrounding a land-locked sea, the Mediterranean. Naturally, Jerusalem was its geographical centre.

Around 1400, a copy of Ptolemy's long forgotten *Geographica* was brought back to Venice from Constantinople. It caused quite a sensation, because it included unknown regions such as the Canary Islands, Ceylon and the Indian Ocean. It even suggested the existence of a territory situated in the Far East. Although Ptolemy's maps later proved to be far from accurate, they spurred interest in cartography and instrument-making.

As such, Columbus's search for Asia (1492) vastly underestimated the earth's circumference; had it been accurately calculated (which the Ancient Greeks had managed to do), Columbus's voyage would not have been financed!

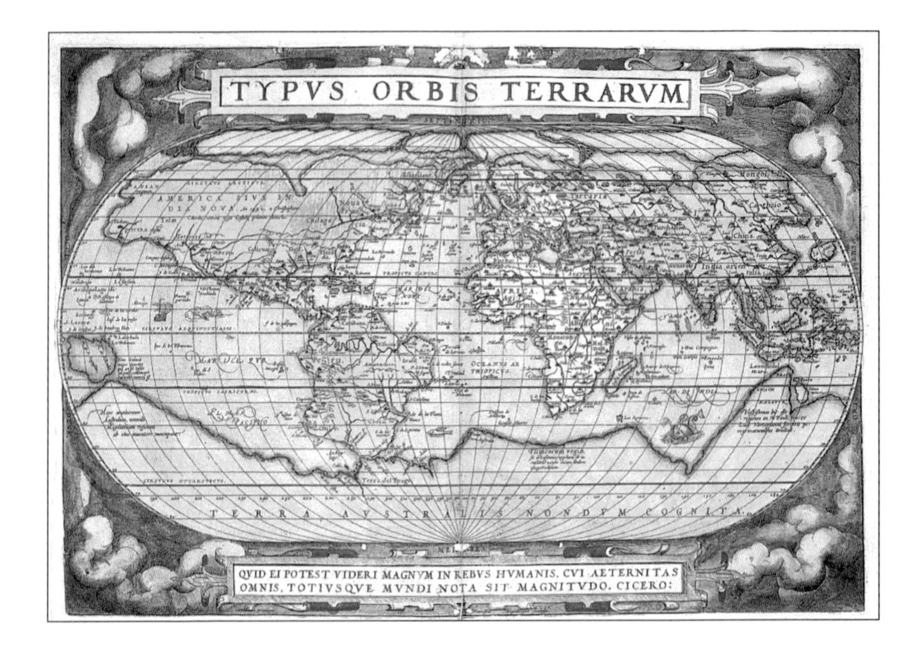
TYPVS ORBIS TERRARVM
TERRA AVSTRALIS NONDVM COGNITA
QVID EI POTEST VIDERI MAGNVM IN REBVS HVMANIS, CVI AETERNITAS OMNIS, TOTIVSQVE MVNDI NOTA SIT MAGNITVDO. CICERO:

WORLD MAP, ABRAHAM ORTELIUS, SIXTEENTH CENTURY

A new science was born: geography. The Flemish cartographer Gerhard Mercator published maps based on his projection system adapted from Antiquity. This method solved the problem of representing a sphere on a flat plane (like an unrolled cylinder). The meridians of longitude are equally spaced lines uniting at the poles, and the latitudes are perpendicular to the meridians. This map suggests that all continents were once united, a theory that was to be confirmed in the twentieth century.

Map C.2.d.7.1-2. By permission of The British Library, London

The astrolabe, known since Antiquity, the needle compass invented by the Chinese, and Renaissance navigational charts were the tools that redesigned the world. These scientific objects, which we consider to be works of art, stimulated advances in ship design. Large, three-masted ships that were easier to manoeuvre in high winds, replaced the medieval cogs. Precious shipments could be transported in safer conditions.

More changes were at hand, however, as material and spiritual concerns tended to coincide. Jerusalem was taken by the Muslims and the Christian world found itself without a centre. This disorienting experience cast doubts on many beliefs and marked the beginning of modern Western history.

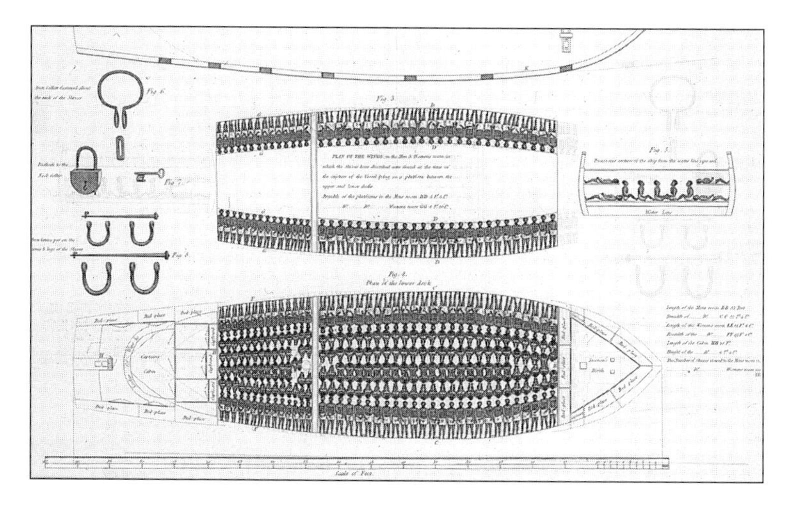

PLAN OF SLAVE SHIP
With advances in ship construction, all the oceans would prove to be connected and no reasonable person could doubt that the earth was anything but round. Yet, science and humanitarian progress were not synonymous . . . since over 10 million people were deported as slaves from Africa to the Americas.
National Maritime Museum, Greenwich, London

ASTROLABE, GUALTERIUS ARSENIUS, 1567
Until the Renaissance, navigation largely remained coastal. The sea astrolabe would give a rough estimate of latitude based on the position of the stars. The magnetic compass would especially help mariners to determine their locations. New instruments for measuring time and distance were developed in emerging disciplines. Navigation, cartography, architecture and engineering demanded apprentices grounded in the principles of perspective.
© Musée des Arts et Métiers-CNAM, Paris. Photo Studio Cnam

Scientific architecture

The age of exploration was literally expanding horizons with each new expedition. Westerners established great confidence in their capacity to change the course of events and developed a new attitude towards knowledge. The centre of the world progressively moved from the Mediterranean region to the North-West of Europe.

Meanwhile, science contributed to an architectural milestone of the same period, the dome of Florence Cathedral, symbol of the 'New Athens'. It was the work of Filippo Brunelleschi (1377–1446), the developer of one-point perspective, an architect and painter who started his career as a sculptor. He had lost an earlier competition for the cathedral baptistry doors to the sculptor Lorenzo Ghiberti (1378–1455), and went to Rome to study ancient monuments.

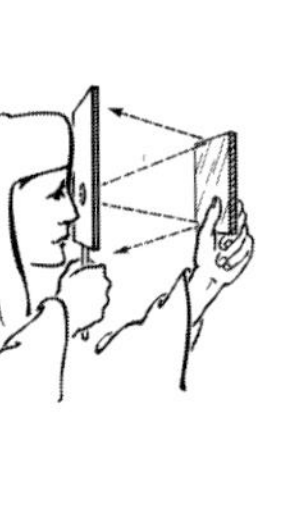

BRUNELLESCHI'S EXPERIMENT, PHILIPPE COMAR

Renaissance architecture began with Brunelleschi, whose development of one-point perspective, for architectural purposes, revolutionized art. Linear perspective assumes that parallel lines, receding from our eyes, converge at a point on the horizon, and that the diminution in size of objects is directly proportional to their distance from us. To demonstrate his method, he set up a mirror facing a building, then painted the reverse mirror image on a flat wooden panel. He made a hole in the centre of the painting. When viewers looked through the hole, the real building could be seen and compared with its image painted in perspective.

La Perspective en Jeu, *Découvertes Gallimard, 1992*

Rome at the time looked like a vast open-air museum whose antique objects were beginning to be appreciated for their aesthetic value (as well as for the commercial value of marble!). Brunelleschi may well have discovered the technique of in-perspective drawing while trying to record accurately the appearance of antique architecture.

He returned to Florence to participate in the contest for designing the cathedral dome – to rest on an octagonal base 40 m wide – and found Ghiberti once again a competitor. But this time, Brunelleschi met the challenge. His solution, a huge ovoid structure made of separate interior and exterior shells, rather than a heavy solid mass, allowed for substantial savings on materials. This construction required precision instruments and, in order to lift the stones into place, Brunelleschi had to invent tools, cranes and derricks. His success rested also on scale drawings and maquettes called *modello*. Although the approach was totally original, the dome consisted of eight curved double panels sustained by ribs that were still reminiscent of the Gothic style.

The break with the medieval trial-and-error custom occurred just before the rebirth of architectural theory. Leon Battista Alberti (1406–72), an architect with a passion for music and mathematics, wrote a ten-volume treatise based on the work of Vitruvius. He wanted to demonstrate that: 'the arithmetic ratios ruling over music and architecture also govern the universe, being of a divine nature.'

This book became the cornerstone of Renaissance architecture and foreshadowed projective geometry. Alberti's rule was: 'It is inconceivable to add or take away any part of (a work of art), without impairing the harmony of the entire structure.'

In the footsteps of his Florentine predecessors, Andrea Palladio (1508–80), initially a sculptor, effected the final rupture with the Gothic style by arguing the case for crisp geometric lines, inspired by an ancient belief in the spiritual value of numerical ratios.

Palladio's buildings were constructed around Venice, a crowded area reflecting its commercial ties with the East. The wealthy sought refuge in the countryside and built themselves neo-classical retreats ornamented with antique sculptures. This architecture displayed the radical change in self-perception then under way as a result of the new economic order. Palladio was the only architect who ever gave his name to a style: his fame crossed the oceans.

The year 1543 was a landmark of the beginning of the scientific revolution; it saw the publication of two theories that irrevocably altered man's relationship to the universe: *On the Revolutions of Celestial Spheres* by the Polish astronomer, Nikolaus Copernicus (1473–1543) and *On the Structure of the Human Body* by the Flemish physician, Andreas Vesalius (1514–64).

Copernicus ousted the earth from its time-honoured position at the centre of the universe by proposing that it revolved around the sun. He wrote: 'And thus rightly in as

FLORENCE CATHEDRAL ILLUMINATED BY LASER AS PART OF *TWO ENVIRONMENTS FOR PEACE*, DANI KARAVAN, 1978
This dome symbolizes the traditional link between advanced technology and architecture. The laser beam, like a blue line of science, was drawn in the Florence sky as a homage to Galileo. The beam departed from the Belvedere fort; from there it travelled towards the cathedral, touching the lantern on top of Brunelleschi's cupola. In order to verify his theories, Galileo had had an opening pierced to let the sunlight through.

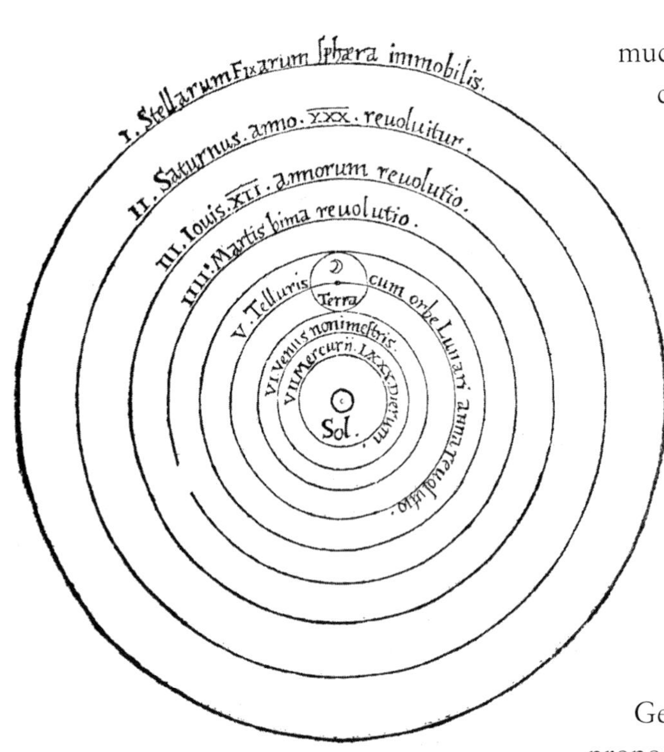

THE UNIVERSE ACCORDING TO COPERNICUS
Ptolemy believed that the earth was circled by the seven known planets (therefore the week is made up of seven days): the Moon and the Sun, Mercury, Venus, Mars, Saturn and Jupiter. Copernicus declared that the sun was the centre of a system around which the earth and the six other planets circle.
Library of Congress, Washington, D.C.

much as the sun, sitting on a royal throne, governs the circumambient family of stars . . . we find therefore under this orderly arrangement, a wonderful symmetry in the universe, and a definite relation of harmony in the motion and magnitude of the orbs.'

A contemporary of Michelangelo and Leonardo, Copernicus was a man of multiple talents, as a self-portrait revealed. Decades later, his brilliant assumption was confirmed with the availability of telescopes. But his scheme, for all its advances, was still rooted in the static and complex cosmogony of Ancient Greece – juxtaposing up to thirty-four crystalline spheres considered to be the most perfect of shapes. Copernicus's theory was simplified by the German astronomer, Johannes Kepler (1571–1630), who proposed fewer orbits that were elliptical rather than circular – implicating an as-yet-unknown force to keep heavenly objects on course. Kepler was inspired by Dürer's treatise on geometry exemplifying that plane sections of circular cones are ellipses. Aesthetics and mysticism played a role in Kepler's theories. Moreover, he considered architecture, music, mathematics and astronomy all to be related.

Astronomy reached a peak with Galileo Galilei (1564–1642), a professor of physics and military engineering who manufactured instruments and researched with rulers and scales, clocks and thermometers. Initially he used the telescope, an invention based on lenses perfected by Dutch opticians, to trace the trajectories of projectiles. This research was of particular interest to the ambitious republic of Venice.

When Galileo turned his telescope towards the sky, he detected the presence of spots on the surface of the moon, as well as previously unnoticed bodies. These irregularities constituted disturbing evidence of the imperfection of the heavens.

Galileo systematically combined observation and measurements with logic and is therefore considered as the father of scientific methodology. He said: 'Philosophy is written in this grand book, the universe, which stands continually open to our gaze, but it cannot be understood unless one first learns to comprehend its language. It is written in the language of mathematics, and its characters are triangles, circles, and other geometric figures, without which one wanders about in a dark labyrinth.'

Called 'the Tuscan artist with his optic glass', Galileo had a lifelong interest in poetry, music and drawing and was admired as a painter in his youth. His friend, the artist Ludovico da Cigoli (1559–1613), said that 'it was art that taught Galileo everything he knew about observation and perspective.'

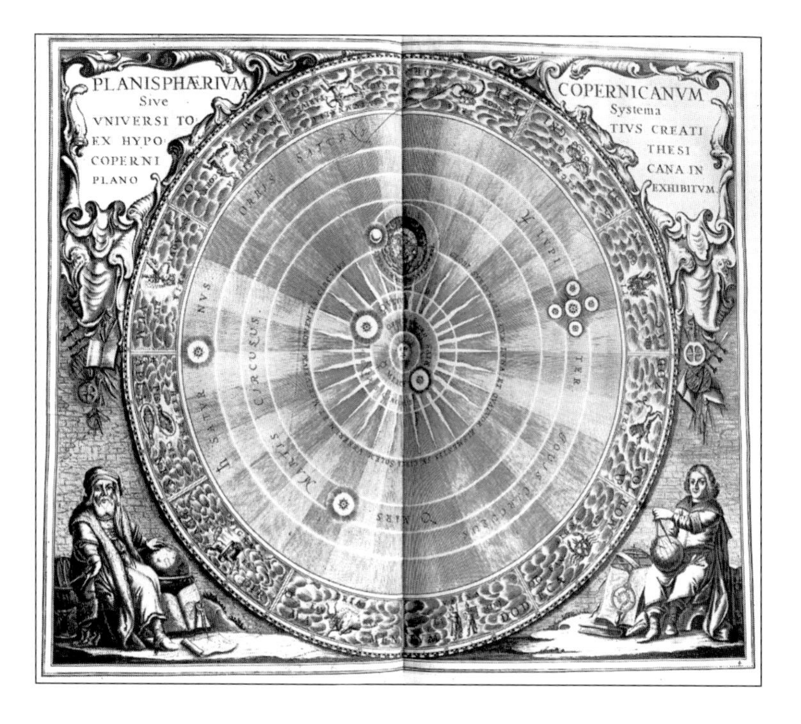

THE SYSTEM OF COPERNICUS, CELLARIUS, SEVENTEENTH CENTURY
Galileo confirmed Copernicus's hypothesis. The four satellites of Jupiter – like a miniature solar system – were also discovered by Galileo using a telescope.
Observatoire de Paris

While astronomers were redesigning the cosmos, other scientists were absorbed by the inner space of the human body. Vesalius, with his pioneering dissections of corpses, opened the way for modern surgery. Influenced by Leonardo, he had his anatomy albums illustrated by a pupil of Titian. Vesalius published his masterpiece at the age of 28. It described the methods and instruments for dissection, together with instructions on how to represent anatomy graphically.

The approval of dissection by Pope Leon X (1475–1521), who posed for Raphael, and Pope Jules II (1443–1513), Michelangelo's protector, brought the practice within the bounds of legal activity.

Modern medicine, with its branches of surgery and pharmacology, was in the making. Parallel discoveries in

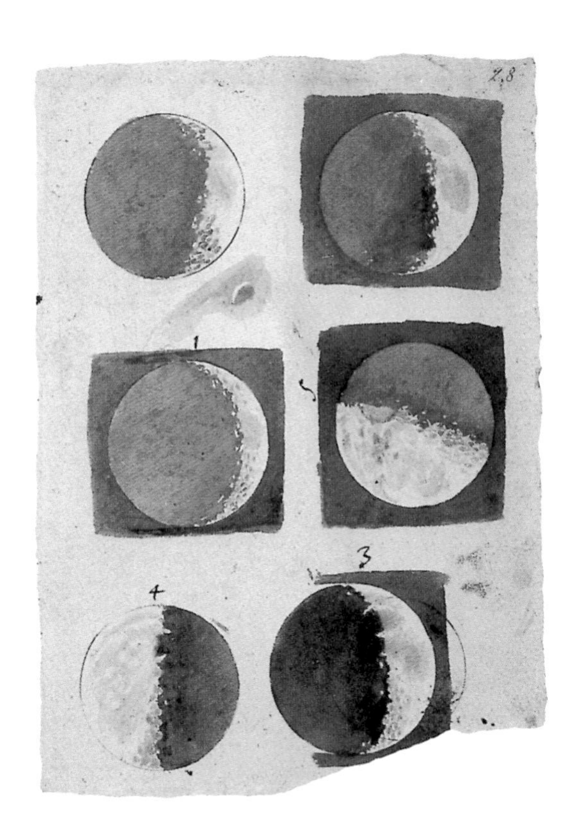

astronomy and medicine led the English physician, William Harvey (1578–1657), to formulate his theory on blood circulation.

About the motion of the blood, he wrote: 'The heart is the beginning of life, the sun of the microcosm; the sun in turn might well be designated the heart of the world; for it is the heart by whose virtue and pulse the blood is moved. The heart like the prince in a kingdom, in whose hands lie the chief and highest authority, rules over all.'

Harvey's poetic description indicated that the human body was still viewed, as in ancient times, as a structural analogue to the universe.

Astronomy and astrology would be influential in medicine well into the following century. But, as the sun found its place in the universe and the heart its place in the body, man reshaped the environment. Planetary motions were now considered to be held in equilibrium by dynamic forces.

Six Phases of the Moon, Galileo, 1616
Galileo trained in drawing and watercolour and enjoyed making scientific illustrations.
Ms.Gal.48c.28r. Biblioteca Nazionale Centrale, Florence

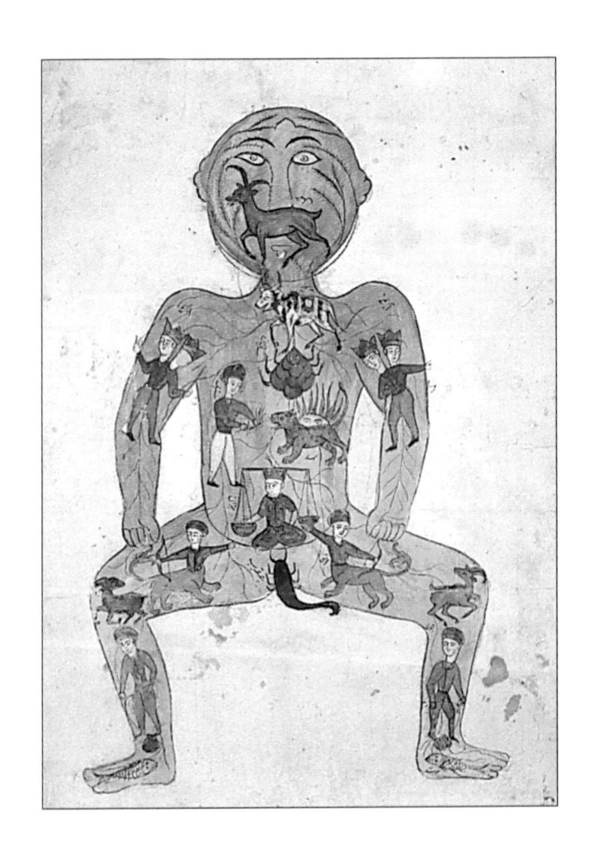

Zodiacal Man, Persian manuscript, sixteenth century
The *Zodiacal Man* shows parts of the body over which particular signs were thought to have powers. Such indications were used for bloodletting. The body was thought to be regulated by the four humors: blood (heat), phlegm (cold), black bile (melancholic) and yellow bile (choleric). These humors were viewed as subdivisions of the general cosmogony; even Galileo had astrological predictions made on his daughter's birth.
Wellcome Institute Library, London

PORTRAIT OF PHILIPPUS PARACELSUS,
PETER PAUL RUBENS, SIXTEENTH CENTURY

The Swiss physician Paracelsus was a forerunner of pharmacology. The 'Luther of chemistry', who changed the emphasis of alchemy from making gold to making medicines, was admired by generations of scholars, painters and poets.

Musées Royaux des Beaux-Arts de Bruxelles.
© A.C.L.-Brussels IRPA-KIK. Photo Cussac

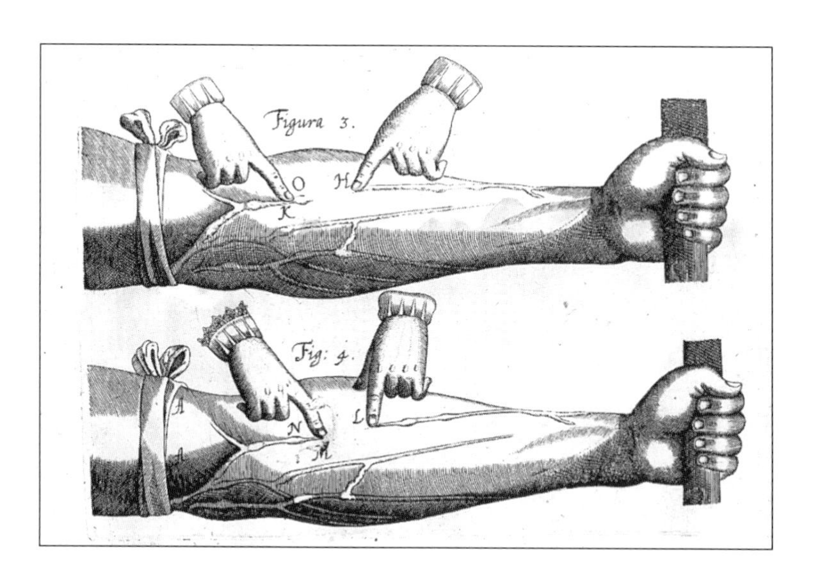

ENGRAVING REPRESENTING THE BLOOD CIRCULATION,
WILLIAM HARVEY, 1628

The demonstration of blood circulation is illustrated here: the arm is bound above the elbow so that the veins stand out (the valves show as swellings). Pressure causes blood to empty out of the vein between the pressure point and the next valve towards the heart.

History & Special Collections Division. Louise M. Darling Biomedical Library, UCLA

In a context of religious turmoil across Europe, the Catholic Church developed a glorious baroque combination of art and science to magnify its power. Painting, sculpture and architecture were produced in order to picture religious miracles. A manual of iconography which codified aesthetic rules was established to this effect.

The baroque style was dominated by Gian Lorenzo Bernini (1598–1680), a sculptor, painter and architect who gave Rome its grandiose imprint. The notion that a hierarchy of angels could no longer propel the planets was illustrated in papal architecture – in which the baroque 'sky/ceiling' designer attempted to compensate for the tension between contradictory visions of the universe.

Baroque appealed to courts throughout the world, including the recently conquered Americas to which architects and artists were sent to build and decorate. The late baroque period

TRIUMPH OF THE DIVINE PROVIDENCE,
PORTION OF A CEILING OF THE PALAZZO BARBERINI,
PIETRO BERRETTINI DA CORTONA, SEVENTEENTH CENTURY

The *Paradise* by Tintoretto (1519–94), a mannerist painter, was described in connection with Copernicus' solar system – a theory probably unknown to the artist. Baroque ceilings, such as this one, were called 'post-Copernican' skies. They represent a universe that is visually integrated into the architecture, in which one can easily imagine magnificent opera performances.

Alinari Anderson-Giraudon

was characterized by an effusion of curvaceous forms and geometric elements designed with the help of the most sophisticated mathematical concepts of the day – to reflect dynamic forces at work. Fascination with the idea of infinity, an all-embracing concept of the unity of all things, dominated this art, which would have been inconceivable without the stimulus of astronomy.

At the same time, between 1650 and 1750, France had become the wealthiest nation in the Western world. At Versailles, landscape was geometrically transformed according to orthogonal rather than circular and elliptical rules. Most modern technologies were applied to buildings, gardens and canals. Kilometres of cast-iron pipes, produced by state-of-the-art methods, supplied water to vast gardens and waterworks. A gigantic pump, the Marly machine, supplied 1,400 fountains.

Art and industry were never closer as a multitude of crafts enhanced architecture. 36,000 people – a workforce comparable with that of imperial Rome – were involved in furniture- and cabinet-making, gold- and silver-smithing, silk- and tapestry-weaving, mirror- and glass-making. Some institutions that were active then – the St Gobain glassworks and the Gobelins tapestries – are still in operation.

In France, baroque design developed but quickly lost ground to rational shapes and thoughts. On his only trip abroad, Bernini's advice to King Louis XIV was not heeded. A sober style was imposed by yet another artist-scientist, Claude Perrault (1613–88). This French architect, physicist and physician built a classic colonnade for the new addition to the Louvre.

Meanwhile, René Descartes (1596–1650) had suggested that God created a world which ran like a huge clock. He had tried to prove God's existence by applying mathematical methods to metaphysical questions. In so doing, Descartes paved the way for Newton.

Astronomers, recognizing the need to identify the force that keeps planets in their orbits, had to wait for the discoveries of Isaac Newton (1642–1727) – born in the year of Galileo's death. In his work *Principia,* Newton explained that the motion of planets around the sun, the odd trajectories of comets and the falling of objects towards the earth are all governed by the laws of universal gravitation.

This theory found ultimate confirmation with the return of Halley's comet seventy-six years later, exactly as predicted.

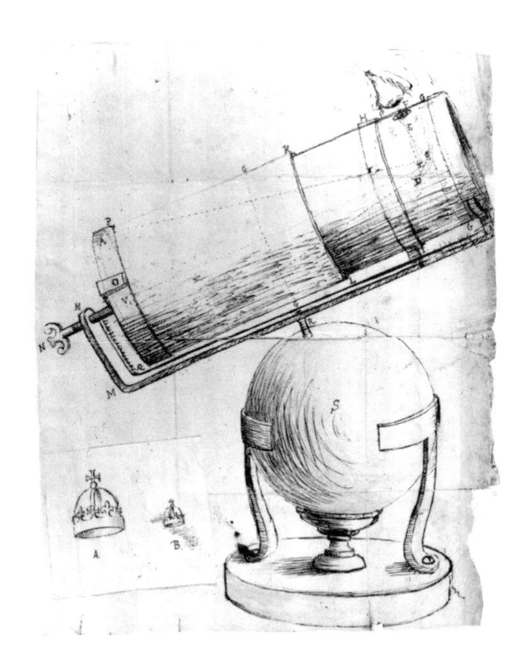

ISAAC NEWTON'S TELESCOPE, SEVENTEENTH CENTURY
Newton made his own mirrors and lenses and developed a telescope containing a special type of mirror still featured in modern instruments.
By permission of the President and Council of the Royal Society, London. © The Royal Society

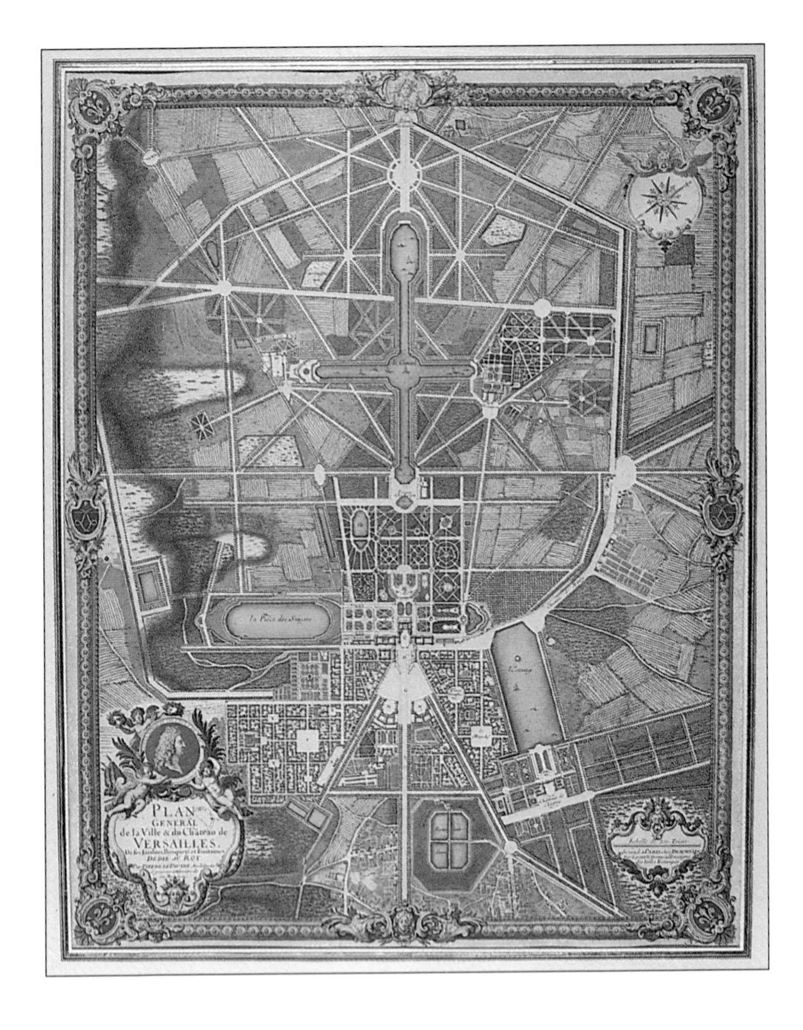

MAP OF THE CITY AND PALACE OF VERSAILLES WITH ITS GARDENS AND FOUNTAINS, PIERRE LE PAULTRE, SEVENTEENTH CENTURY

In France, formal geometry was preferred to baroque convoluted rhythms. Perfectly landscaped parks were correlated with palace construction and city streets. André Le Nôtre designed the gardens of Versailles, as well as the Champs Elysées.

Bibliothèque Municipale de Versailles. Photo Jean Vigne

(For good reasons, Halley had published the first edition of Newton's *Principia* at his own expense.)

Newton's approach to elucidate the mechanical wonders of nature was essentially mathematical; it included inductive and deductive stages. Weight, volume, temperature and other parameters could be determined with a higher degree of accuracy by his time. Instruments originally designed as tools for observation were becoming experimental tools.

It was then that physics split off from metaphysics, which was not an easy parting, as Newton himself respected authority on all matters, even when it contradicted science. 'This most beautiful system of the sun, planets and comets could only proceed from the counsel and dominion of an intelligent and powerful Being . . . (God) endures for ever and is everywhere present, he constitutes duration and space.'

Newton spent a major part of his life researching the secrets of alchemy (without publishing this work). He implicitly recognized that his creative approach was partly irrational.

Newton said: 'I do know what I may appear to the world, but to myself I seem to have been only a boy playing on a sea-shore, and diverting myself in now and then finding a smoother pebble or a prettier shell than ordinary, whilst the great ocean of truth lay all undiscovered before me.'

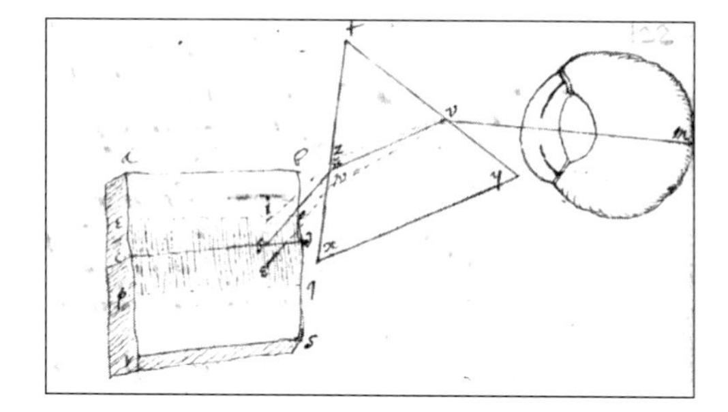

Decomposition of white light through a prism, Isaac Newton, seventeenth century
Having observed white light crossing a rainbow, Newton analysed it as it passed through a prism and broke up into its colour constituents. He showed that, when the different components were fed back through a second prism, white light was reconstituted. Newton's *Optics* were subject to debate by scientists and artists alike.
f.122r manuscript Add.3996. By permission of the Syndics of Cambridge University Library

He was fascinated by visual phenomena and set the foundation of modern optics. He also took great pleasure in illustrating his notebooks. Einstein said of him: 'In one person he combined the experimenter, the theorist, the mechanic, and not least, the artist. . . . His joy in creation and his minute precision are evident in every work and every figure.'

The dawning of the mechanical era marked the advent of a split between art and science. Newton's strongest supporter, the French philosopher Voltaire (1694–1778), commented in a cautionary tone: 'Everyone works at geometry and physics . . . sentiments, imagination and the finer arts are banished. Not that I am angry that science is being cultivated, but I do not want it to become a tyrant that excludes all else.'

Royal Observatory from Crooms Hill, English School, eighteenth century
Astronomical and meteorological measurements were made from the room at the top of Christopher Wren's observatory in Greenwich. This location became the official prime meridian geographical reference two hundred years later.
National Maritime Museum, Greenwich, London

A new concept: progress

London was entirely rebuilt after the great fire of 1666. The architect, Christopher Wren (1632–1723), an intellectual prodigy and esteemed friend of Newton, reshaped fifty-one of the city's eighty-seven damaged churches, and redesigned Saint Paul's Cathedral.

Until the age of 30, however, Wren had concentrated on mathematics, physics, astronomy and other sciences, rather than architecture. A close counterpart of the Renaissance artist-scientist, he was, as England's arbiter of architectural taste, the Anglo-Saxon Bernini. He imposed a style for a new age, aptly named Baroque Classicism.

The Industrial Revolution was on its way. Deforestation led to the exploitation of coal as the main source of energy. The United Kingdom, prospering from the wealth extracted from its expanding empire, produced more than 80 per cent of the world's coal, used in the manufacture of new materials such as glass and chemicals (detergents, bleaching agents, etc.)

Coal was also used to fuel furnaces instrumental in the production of cast iron. Coal-mining, in turn, required cast-iron steam engines to pump water out of mines, so that more coal could be mined from deeper levels. This cycle accelerated the shift to a machine-dominated economy.

Until then, a few simple objects – lever, pulley, wheel, axle, screw and hammer – had remained the basic tools. A flood of new inventions now led to belief in the inevitability of progress. Building materials and techniques were developed in parallel: iron and bricks replaced wood. Industrial chimneys defined the landscape, while a residential building style flourished.

THE HUMAN BRAIN, CHRISTOPHER WREN, IN *CEREBRI ANATOME* BY THOMAS WILLIS, 1664
Sir Christopher Wren, who changed the face of England's built landscape, excelled at anatomical drawings.

Bodleian Library, University of Oxford. Shelfmark = Lister B.66. Between pages 12-13. Photo Bodleian Library, Oxford, UK. Copyright

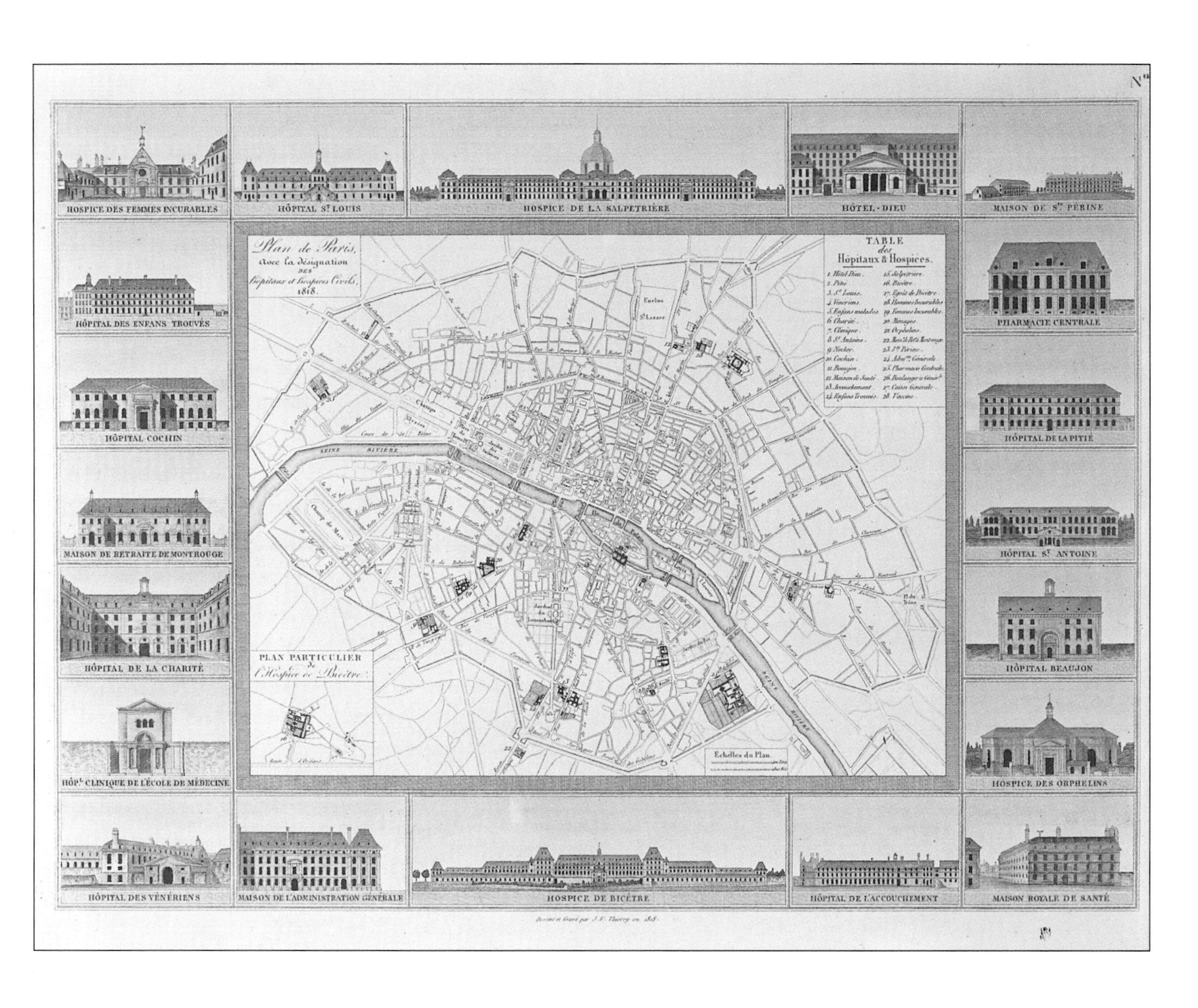

PARIS PUBLIC HOSPITALS AND HOSPICES, 1820

At the beginning of the nineteenth century, Paris was the world centre for medicine. The city's abundance of hospital beds – there were nearly 40,000 – was a by-product of France's violent history: the Revolution and its aftermath, Napoleon's military campaigns.

Musée de l'Assistance Publique. Photo J.-L. Charmet

MONTICELLO, THOMAS JEFFERSON'S RESIDENCE, UNITED STATES, EIGHTEENTH CENTURY
Jefferson, who played a major role in American architecture, had his tomb inscribed with the following epitaph: 'Here was buried Thomas Jefferson, author of the *Declaration of American Independence,* of the *Statute of Virginia for Religious Freedom* and Father of the University of Virginia' and, as he requested, 'not a word more', although the list of his achievements could fill a book.
Photo UNESCO/L. Alberto

Architectural concepts could be said to follow a straight line, drawn from Palladio through England, on to a New World inspired by a different order. The American President, Thomas Jefferson (1743–1826), a scientist, agronomist, linguist and educator, was also a town planner and architect who synthesized continental, colonial and antique style in a manner that suited the land of opportunity.

The 'Academic village', as Jefferson called his university in Virginia, exemplified this style. To house the ever-increasing flood of information, architectural types appeared, such as libraries that looked like temples dedicated to knowledge.

Architects in Europe aspired to innovation. For example, Étienne-Louis Boullée (1728–99) designed visionary projects for hospitals and factories that still surprise us by their embellished geometrical abstraction. But social unrest and war placed food supply ahead of architecture on the list of priorities.

Unremitting cycles of destruction and reconstruction created a considerable need for experts. Moreover, invention patents were established in order to abolish the nobility's privileges, and applied research was greatly encouraged (for example, preserving food in cans was an idea that originated from the necessity to supply armed forces).

Engineering moved forward with the creation of institutions such as the École Polytechnique in France, and the Massachusetts Institute of Technology. To this day, they have an impact on materials and thus on building design.

Driven by technical developments in manufacturing and transportation, the Industrial Revolution had a great effect on scientific theory. For example, the study of friction of machine parts led to the understanding of energy conversion, and paved the way for modern physics.

Electricity, a research subject largely spurred by military and meteorological concerns, became a major focus of investigation. The American scientist and statesman Benjamin Franklin (1706–90) showed that an electric force could magnetize and demagnetize iron rods, and protect from lightning. He thus proved by flying a kite in a thunderstorm – and by the shocks he might have received – that lightning was of an electrical nature.

Realizing that electricity was not just a static force, but that metal could conduct it, he triggered an avalanche of experiments on both sides of the Atlantic. Franklin, also a master printer, a moralist and diplomat, was instrumental in establishing the American Philosophical Society, the United States' first scientific society. He truly symbolizes 'the American success story'. Franklin was one of the statesmen involved in the drafting of the Declaration of Independence on 4 July 1776. In France, he was popularly nicknamed *l'ambassadeur électrique.*

Electricity and magnetism were rapidly put to practical use by Michael Faraday (1791–1867), an English autodidact who conceived the electric motor whose energy could be transported and used at a distance; within ten years, it was ubiquitous.

Faraday truly operated the switch from the mechanical to the electrical era, formulating electricity's basic principles with concepts such as ions, anodes and cathodes. The visionary implication of his work was that electricity had to be composed of elementary particles – electrons, which were identified much later. Einstein kept a portrait of Faraday, whom he regarded, together with the Scottish physicist James Clerk Maxwell (1831–79), as the founder of modern physics.

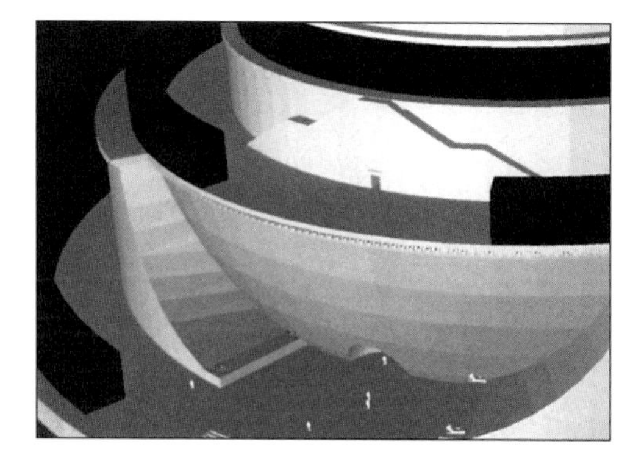

NEWTON'S CENOTAPH ACCORDING TO ÉTIENNE-LOUIS BOULLÉE, COMPUTER RECONSTRUCTION BY JEAN-LOUIS SCHULMANN, 1984

Nearly a hundred years after Newton's death, his mechanical view of the universe had won out over all others. This burial monument – a gigantic hollow sphere – had to be of a magnitude commensurate with the scientist's fame. The tiny white figures give us an idea of the architect's ambition. Boullée's unbuilt megaprojects continue to fascinate to this day.

© J.-L. Schulmann/Image Digitale

'LIGHTNING CONDUCTOR FASHION', C. 1778

Among Franklin's many inventions, the lightning rod launched a hat craze. He invented bifocal glasses and the Franklin stove, which has not been greatly improved upon since.

Ann Ronan Picture Library/Image Select

The nineteenth century ushered in an era of urban expansion. A few decades after the appearance of iron for industrial purposes, such as railways, it was introduced as a building material for its strength and malleability. Soon, iron columns and arches were the standard means for supporting roofs over large spaces: train stations, shopping arcades and exhibition halls.

Architects smitten by technology hastened to build in a welter of neo-Renaissance and neo-Gothic styles, designing to suit new materials – iron and glass – rather than choosing the material best suited to the design.

Yet, a few, notably the Spaniard, Antonio Gaudí (1852–1926) and the Frenchman, Gustave Eiffel (1832–1923), conceived original shapes. The preparatory studies for *La Sagrada Familia,* Gaudí's unfinished cathedral, indicate that his forms are based on mathematics and mysticism.

The century was crowned by the tallest free-standing structure ever built, the Eiffel Tower (1887–89). The architect of the Paris opera house, Charles Garnier (1825–98), considered it a monstrosity and went so far as to circulate a petition demanding that it be demolished (as planned after the World Exhibition of 1900).

Iron technology and precision engineering advanced rapidly. Parts had to be standardized and the need for suitable materials grew: wrought iron (which contains no carbon) for making

Plan of Paris indicating the interventions of Baron Georges Eugène Haussmann from 1853 to 1870

Urban order emerged as a consequence of increasing traffic, which would greatly inspire Impressionist and Futurist painters. New streets are shown by heavy black lines, and new districts by cross-hatching. The areas covered with plant patterns are the great parks, the Bois de Boulogne in the west and the Bois de Vincennes in the east. Radical transformations were similarly effected in Brussels and Vienna, despite the fact that they caused social upheaval.

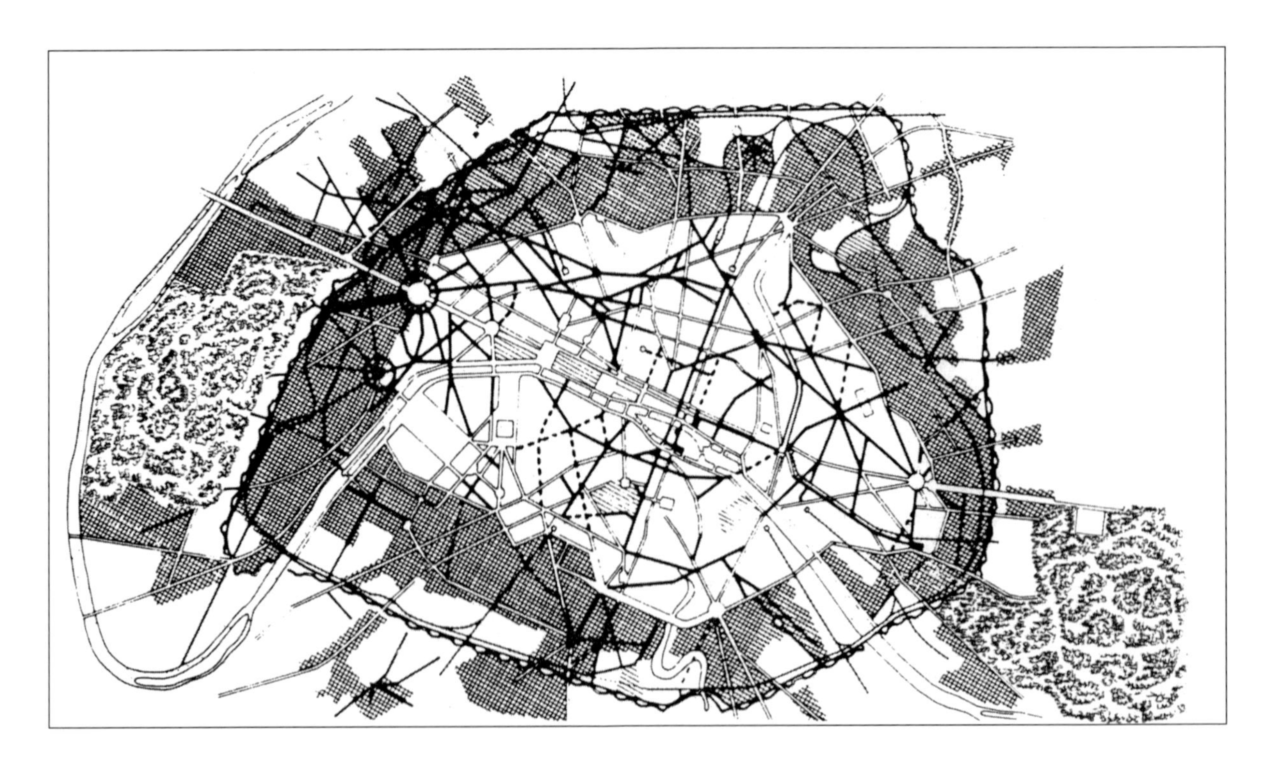

moulds, was too soft for use in rapidly moving machinery, and cast iron (which contains a high amount of carbon) for cutting tools, was too brittle. At last, steel, a light, durable and flexible material (with an average amount of carbon), replaced iron construction.

Steel revolutionized Western architecture as radically as concrete had in Roman days, and the two materials were now combined as reinforced concrete, in which steel rods were embedded. This hybrid material, when used together with a steel framework, resulted in taller and less bulky buildings than ever before.

Architectural experiments were pioneered, especially in the United States. During the century that saw the beginning of democratic urbanism, cities were being designed on the basis of such concepts as traffic circulation. Specialists of all sorts proliferated: urban planners, electricians, heating and

THE BROOKLYN BRIDGE, 1883
Innovative techniques were used in the construction of the elegant neo-Gothic Brooklyn Bridge: a revolutionary method of spinning cable was developed for suspension.
Photo UNESCO/N. Levinthal

ventilation experts, etc. After fire destroyed a large part of Chicago in 1871, reconstruction provided unprecedented opportunities. The rebuilt city, called 'electricity city, or city of speed', delivered the first skyscraper, which was ten storeys high. Elisha Graves Otis (1811–61) then made elevators safe by inventing a device that prevented them from falling if the cable broke.

By 1931, a modern cathedral, the 102-storey Empire State Building, was completed within less than two years, thanks to tubular steel scaffolding, steam shovels and hydraulic jacks.

Skyscrapers soon defined the horizons of fast-growing cities. This architectural form developed in three overlapping phases;

an introductory classical style, an ornate theatrical one, and a more lasting international style. They became laboratories for new forms of technology permeating all aspects of modern life.

Thomas Edison (1847–1931), the father of modern R&D (organized research and development) – together with his fifty-strong team – patented over 1,000 inventions. The self-made man, whose formal education consisted of just a few years in public school, was constantly drumming up backing for his 'research factory'. He never developed an idea without a prior market study. Of Edison, Henry Ford said: 'He definitely ended the distinction between the theoretical and the practical man of science, so that today we think of scientific discoveries in connection with their possible present or future application to the needs of man.'

Inventor-entrepreneurs such as Alexander Graham Bell (1847–1922), who popularized the telephone, and George Eastman (1854–1932), whose photography firm introduced the first employee profit-sharing system, set the foundations of an avant-garde who applied inventions to purposes extending far beyond people's wildest dreams.

MODEL FOR A *MONUMENT TO THE GLORY OF THE THIRD INTERNATIONAL IN MOSCOW*, AFTER VLADIMIR TATLIN, 1919

This structure conceived by the Russian father of Constructivism, Vladimir Tatlin, was meant to be twice as big as the Empire State Building. A contemporary critic described it in the following terms: 'The monument consists of three great rooms of glass, erected with the help of a complicated system of vertical pillars and spirals. These rooms are placed on top of each other, and have different, harmonically corresponding forms. They are to be able to move at different speeds by means of a special mechanism. The lower storey, which is cubic in form, rotates around its own axis at a rate of one revolution per year. . .'.

Collections Mnam/Cci/Centre Georges-Pompidou, Paris.
Photo: Photothèque des Collections du Mnam/Cci.
Centre Georges-Pompidou, Paris

Meanwhile, the Swede, Alfred Nobel (1833–96), invented dynamite, which facilitated tunnel construction and thus the expansion of railways. The 'richest vagabond in Europe' created prizes for those who contributed to the well-being of humankind by research or their strivings for world peace. Since Nobel's initiative, science has been increasingly supported by industrial companies that recognize its economic potential.

Comparable investments in art became rare, if in fact they existed at all. Artists started to rebuild their links with society slowly; only recently have galleries, museums and corporations, the modern patrons, started to appreciate art's 'Big Business' potential.

The professionalism code was then set for a new type of individual, the entrepreneur. The twentieth century also saw women, finally, participate in most sectors of activity.

THOMAS ALVA EDISON IN HIS LABORATORY,
W. K. L. DICKSON

Gelatin silver print. Courtesy George Eastman House

Progress in hygiene and medicine shaped modern individuals. While buildings were getting taller and taller, scientists were moving deeper into body organs, tissues and cells. Until Louis Pasteur (1822–95), the founder of microbiology, identified germs as the agents of illness, disease was still believed to be caused by an imbalance of the humors.

By showing that illness was not a spontaneously occurring evil, but that it had a cause that could be eliminated, Pasteur broke the bond that linked man and fate. Thanks to the development of vaccines, the incidence of the great child-killer diseases was greatly reduced. Pasteur also made contributions in fields that represented major economic disasters, such as silk and wine diseases.

Yet, until the age of 20, Pasteur was mostly involved in drawing and painting and, according to his contemporaries, could have become a professional artist. Years later, as the holder of the Chair in applied physical chemistry at the École des Beaux-Arts, he confirmed his dual interest: 'In certain circumstances, I can clearly see the possible and desirable alliance between science and art, and how scientists can take their place next to artists.'

THE MOTHER BY LOUIS PASTEUR, NINETEENTH CENTURY
Pasteur had a well-trained eye. He was the first to notice that crystalline shapes – which had been examined by numerous researchers before him – existed in three forms: right-hand, left-hand and symmetrical shapes. His discovery gave rise to the birth of theory that wended its way into modern art (Robert Smithson's for example), while being essential to the study of molecular structures.
Institut Pasteur

Around the same time, the perception of the self was transformed by Sigmund Freud (1856–1939), the Austrian physician who invented 'archaeology of the mind', so to speak. Before psychiatry was formulated into a theory, the commonly diagnosed emotional ailment was hysteria or a vague state of hypochondria.

Freud's breakthrough, within the singular domain he chose to concentrate on, ultimately spoke to an amazing range of interests and values, especially art. He had a lifelong passion for ancient art and, after his death, his ashes were placed in one of the antique vases he cherished so much.

Both Pasteur and Freud made bold demonstrations that troubling mental and physical conditions could be identified and rooted out. The rapport between humans and the forces of the universe was radically changed once again, bringing about a different cognitive order. A broader concept, which extended to disciplines such as anthropology and sociology, was in the making.

offene Kette. **geschlossene Kette.**

Diese Ansicht über die Constitution der aus sechs Kohlenstoffatomen bestehenden, geschlossenen Kette wird vielleicht noch deutlicher wiedergegeben durch folgende graphische Formel, in welcher die Kohlenstoffatome rund und die vier Verwandtschaftseinheiten jedes Atomes durch vier von ihm auslaufende Linien dargestellt sind:

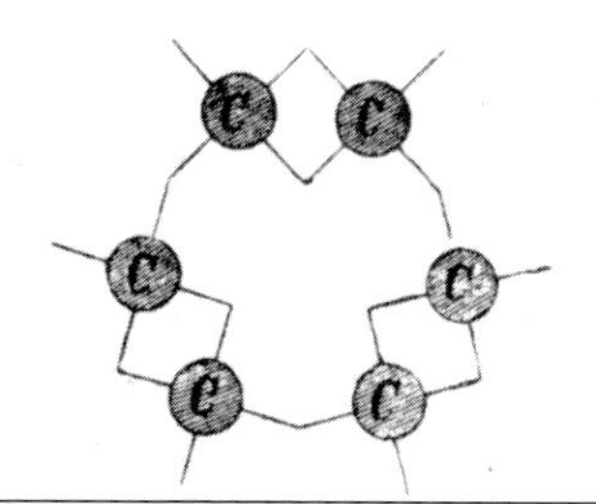

BENZENE RING, 1865
FRIEDRICH AUGUST KEKULÉ VON STRADONITZ
Kekulé, a chemist and one-time architecture student, discovered the hexagonal structure of benzene. Obtained from coal tar and oil, it is used to synthesize numerous products (dyes, medications, plastics, etc.). Legend has it, probably influenced by psychoanalysis, that Kekulé's vision of 'whirling atoms came to him while dreaming in front of the fireplace'.
Organic Chemistry

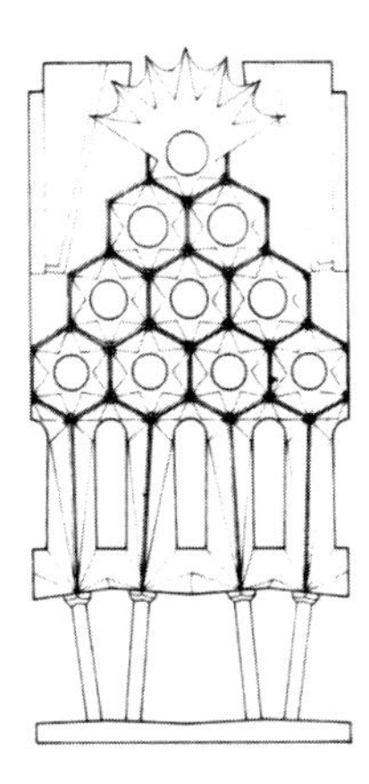

ARCHITECTURAL SECTION OF THE CATHEDRAL
LA SAGRADA FAMILIA BY ANTONIO GAUDÍ

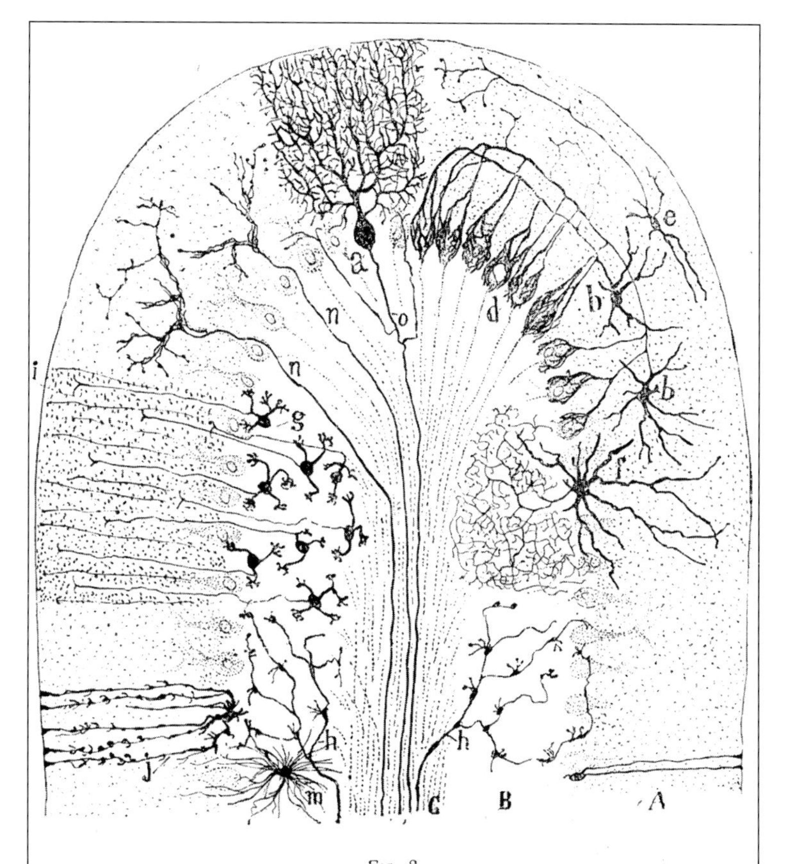

FIG. 8.
Coupe transversale demi-schématique d'une circonvolution cérébelleuse de mammifère.

BRAIN NERVES, SANTIAGO RAMÓN Y CAJAL, 1894
Cajal, who was awarded the Nobel Prize for Medicine – and whose artistic vocation was discouraged by his parents – used the newly discovered dyes and cell-staining techniques to draw the complex interrelations of nerve endings. These 'butterflies of the soul' are described by neurobiologist Jean-Pierre Changeux as: 'Billions of neuronal spider webs in which myriads of electrical impulses crackle, relayed here and there by a rich palette of chemical signals.'
History & Special Collections Division.
Louise M. Darling Biomedical Library, UCLA

Back to the future?

At the dawn of the twentieth century, the discovery of X-rays by Wilhelm Roentgen (1845–1923) generated tremendous public excitement. It furthered the understanding of the atom, which was found to be divisible – yet another revolution.

This called for the redefinition of energy versus matter. Albert Einstein (1879–1955) established the concept of the equivalence of energy (E) and mass (m), in his famous equation $E = mc^2$ (c is the speed of light). His theory of relativity discredited the idea of a universal framework, observations being relative not only to the object's position but to the observer's as well. He indicated that Newton's gravitation could be thought of as a distortion of space.

Like most creative artworks, Einstein's revolutionary science was not understood immediately, even by scientists. But, having predicted a deflection around an eclipse that took place in 1919, he was suddenly hailed as the twentieth-century Newton.

The public had to see proof to believe it. In the demonstration, communication and acceptance of abstract concepts, be it in science or in art, the visual process seems to be essential.

MARIE CURIE IN HER LABORATORY, 1921
Pierre and Marie Curie pioneered the field of radioactivity, for which they received a Nobel prize. Marie pursued research with her daughter Irène and son-in-law Frédéric Joliot, for which they received another Nobel award. Marie spent the First World War at the front lines, using X-rays for medical purposes; she developed radiation sickness and died of leukaemia. The destructive power of the atomic bomb, which she would have deplored, would not be unleashed for another ten years.
Association Curie et Joliot-Curie

Einstein worked primarily by developing his intuition, using reasoning as well as visual-spatial imagination, and said: 'To visualize a theory, or bring it home to one's mind, therefore means to give a representation to that abundance of experiences for which the theory supplies the schematic arrangement.'

The historian George Sarton (1884–1956) said: 'Science always was revolutionary and heterodox; it is its very essence to be so; it ceases to be only when it is asleep.' Then, unexpectedly, modern physics provided the basis for radical change in the study of artworks.

Gamma rays	→	*disinfection*
Ultraviolet and X-rays	→	*authentification, restoration*
Visible spectrum	→	*photography, cinema, television*
Radio waves	→	*sound transmission*
Laser	→	*cleaning sculptures, holography*
Fluorescence	→	*identification*
Radioactivity	→	*carbon-14 dating*

PHYSICS APPLIED TO ART
Several techniques used in science are equally important in the artistic domain. AGLAE, the particle accelerator of the Grand Louvre, is used for non-destructive analysis of precious objects, with a precision of one part in a million.

The implications of the atomic theory would affect art in its inspiration, production and conservation, as well as most sciences: physics, medicine, chemistry and geology. Ultimately the earth itself was affected, since these findings 'culminated' with the nuclear weapon used in 1945.

Einstein was a dedicated peace activist founding his own harmony in: 'music which has no effect on research work, but both are born of the same source and complement each other through the satisfaction they bestow.'

While Einstein initially assumed that the universe was stationary, the American Edwin Hubble (1889–1953) established that the further away a galaxy is from the earth, the faster it recedes. This finding supported the notion of a continuously expanding universe, a theory first hypothesized by the Belgian priest Georges Lemaître (1894–1966). The latter referred to an original 'primordial atom', or a universe that originated with a Big Bang. After deliberately ignoring this theory for a long time, one day Einstein jumped to his feet applauding: 'This is the most beautiful and satisfactory explanation of creation.'

While looking for the infinitely distant, scientists also kept searching for the infinitely small and, in so doing, they tracked down a universal transmitter of inherited characteristics, the gene.

The story of genetics was enacted by three major players, the 3 Ms: Gregor Mendel (1822–84), an Austrian priest who enunciated the basic principles, Thomas Hunt Morgan (1866–1945) who identified the location of genetic material in structures called chromosomes. The third, Barbara McClintock (1902–92), described how genetic material controls the development of cells. Genetics has become the emblem of our age.

Meanwhile, architectural landmarks appeared as industrial empires became their models. The Bauhaus school of design (1920) dedicated its entire effort to integrating art with practical applications derived from science. Emphasis was placed on pleasing design for the humblest functional objects and on mass production of good design.

The aspiration to return to mankind's roots grew in reaction to a machine-dictated lifestyle and building forms that were closer to nature became fashionable: English cottages or Japanese pavilions.

The Chicago architect Frank Lloyd Wright (1867–1959) introduced an artistic symbiosis of light, space and nature, while the Swiss-French architect Le Corbusier (1887–1965) created harmony through a system of proportions based on the human body, in relation to the Golden Section – which he named 'modulor'. He said: 'Behind the wall, the Gods play with numbers, of which the universe is made up.'

SKYLIGHT OF THE ENTRANCE HALL OF A FACTORY, FRANKFURT, 1920
Industrial buildings became temples of avant-garde architecture. The corporate culture has developed thanks to logos and product designs.
© Photo Siegfried Lavda/Hoechst AG

Genes can now be visualized and manipulated into creating new species and to cure diseases: and the first test-tube baby is over 20 years of age. Whether we like it or not, it seems that 'progress' cannot be stopped. Genetics poses many ethical questions.

The search for a better grasp of the structure of atoms through particle accelerators and nanotechnology, together with the quest for infinite space, is urging us to further understand our own function – in an environment increasingly driven by nuclear energy.

Our constructions that need to sustain the delivery of food and thought to the growing masses, are oil refineries, nuclear and surplus processing plants, of which the Paris Centre Georges Pompidou or the Sydney Opera House are noble versions.

From temples dedicated to astronomy, we have moved to designing hi-tech space stations and cyberspace habitacles. What will homes of the future look like? Single-person movable vessels equipped with cordless computers, and appliances integrated into walls of working/living rooms?

Almost anything is conceivable, given that we can assemble architecture in outer space, sending instructions by means of computers from our fragile planet earth. Such futuristic techniques and materials also foster changes in art, particularly in architecture.

VITRY 2001, PHILIPPE STARCK, 1994
This facility, which might incinerate 500,000 tons of material per year, could be functional near Paris by the year 2002. The designer has instilled an unexpected spiritual elegance into this structure's clean lines, reminiscent of antique temples.
Photo Hervé Termisien

Interior view of the Guggenheim Museum designed by Frank Lloyd Wright, Mikiko Noguchi

This architecture echoes a form of archaic biology. Alexander Calder, a sculptor and engineer, dreamed of art that would reflect upon science and used gravity to balance and animate his graceful works. He considered that a composition is defined by the disparity of form, size, weight and movement. 'It is the apparent rupture of regularity mastered by the artist that makes or breaks a work of art.'

Factories and research laboratories continue to reflect a link with science and spirituality. Buildings project human values, whether related to mythology, religion, philosophy, art or science. The myth of a 'cosmic egg' seems as close to the genesis than the laser to the thunderbolt . . . or the chromosome to the chicken egg.

Humanity has considerably enlarged its experience of time and space: we are better equipped than the Ancient Greeks to express analogies between the environment and our internal universe. We nevertheless remain close to them when we create objects and buildings reflecting our sense of a universal order.

Over thousands of years, architecture has developed under the influence of practical necessity: the technology push combined with the architect's visionary pull. Famous builders happened to be brilliant scientists. Imhotep, the so-called god of medicine who built at Saqarra, Anthemius, the geometer who designed the dome of Hagia Sophia, Christopher Wren, the astronomer-mathematician who gave London its new look, all functioned as both artists and scientists. Cross-disciplinary talents such as these can hardly be viewed as simply fortuitous.

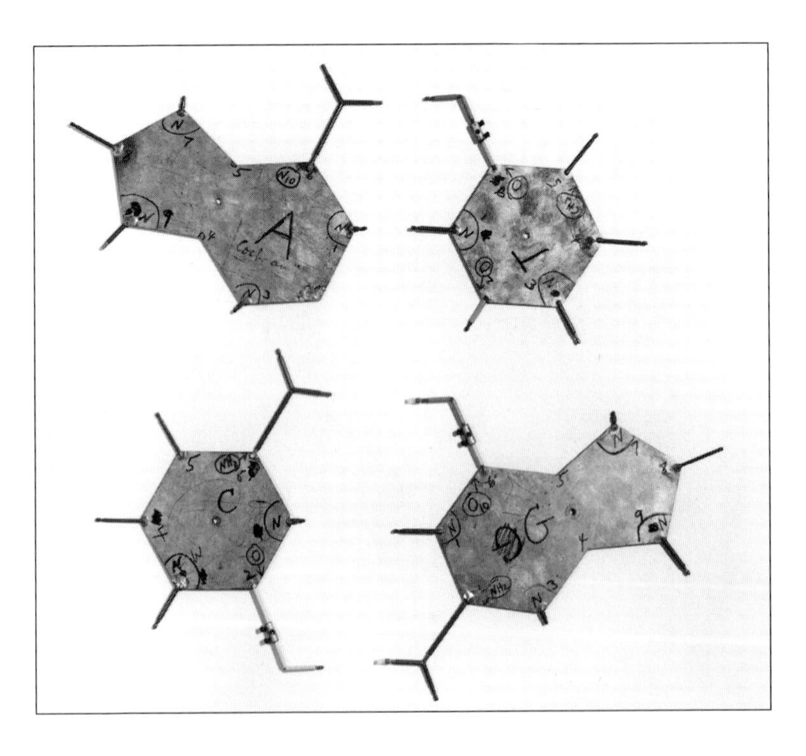

Metal plates used for studying the structure of DNA

These small metal plaques – which one might easily take for Calder mobiles – were used to visualize the structure of the double helix, the repository of genetic material known as DNA.

Science Museum, London.
(Source) Science & Society Picture Library

Double helix model of DNA,
Francis Crick and James Watson, 1953

Science Museum, London.
(Source) Science & Society Picture Library

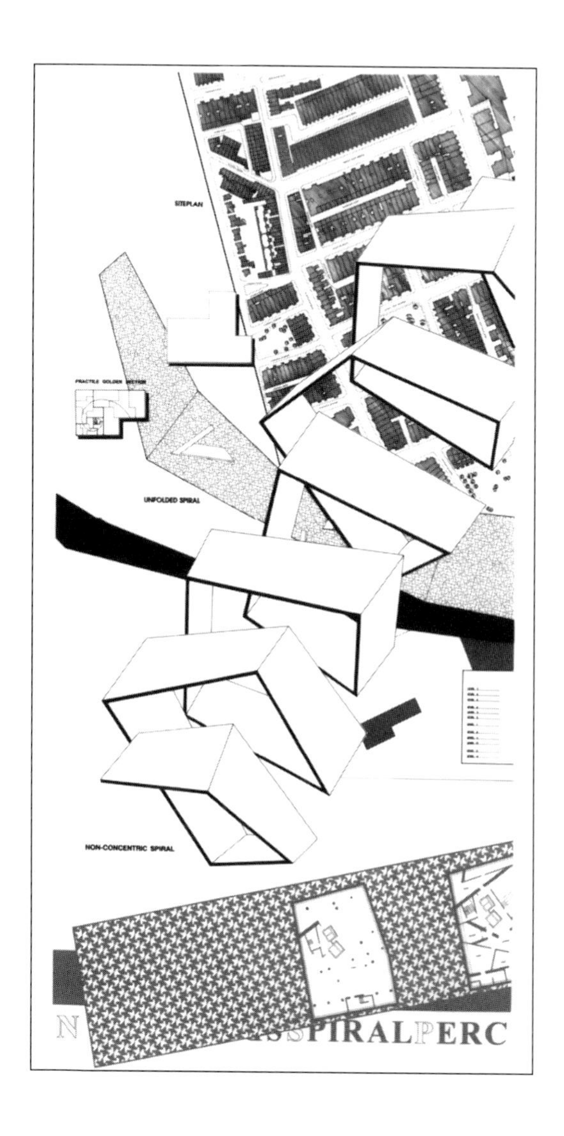

Spiral, architectural project for the
Victoria and Albert Museum in London,
Daniel Libeskind, 1998

© Studio Libeskind

Decoration: a path to hi-tech

FULLER DOME STRUCTURE. *C-60 SOCCER BALL,* KENNETH SNELSON (INSET)
Scientists and artists come up with similar schemes. Fullerene is a molecule made up of a network of sixty carbon atoms, with particularly interesting conductive properties. It was named after the architect Richard Buckminster Fuller, who designed the geodesic dome: a structure that can serve as a roof over a vast space.

Decorative patterns and mathematics

Unlike architecture and the performing arts, which take many of their cues from science and technology, in decoration artists have traditionally led the way. For thousands of years, designs were woven into fabrics, painted on walls and incorporated into the mosaic floors of palaces and temples. One wonders what interior force drove people to cover vast spaces with flowers, spirals and geometrical shapes.

A natural tendency towards simple schemes, such as symmetrical ones, seem to almost respond to a 'biological sense of order'. Moreover, similarities can be found in patterns created independently by artists in Arabic, Christian and Far-Eastern cultures. Perhaps the human brain has some special affinity for repeating patterns.

Ornamentation, reflecting the collective psyche, has reached amazing heights in nearly every culture, from African textiles to Maori tattoos. What would St Peter's Cathedral in Rome or the Palace of Versailles look like, if they were stripped of their decor?

It is only at the turn of the twentieth century that decoration moved backwards in the hierarchy of Western arts. Previously the craftsman's domain, decoration has nevertheless become the designer's prerogative. And, while some would like to ignore it, decor ascends to the level of an autonomous art form – via abstract art.

Throughout history, mathematics formed the basis of decorative patterns. While geometric representation is of a visual order, arithmetic appears to be more closely related to music. The writing of numerals seems to have preceded that of words. The decimal system, based on ten independent signs, was initially developed to enhance the beauty of Indian legends, but its practical aspect would be transmitted to the rest of the world.

Mesopotamians largely laid the foundations for calculation by developing the technique of expressing numbers by position and sign. Ancient peoples in the Middle East worked out practical tables for multiplication and square roots. By contrast, Greek philosophers reached for an aesthetic apogee with conic sections, ellipses and perspective geometry. Aristotle held that: 'Philosophers who believe that there is no place for Beauty in mathematical sciences are wrong. Mathematics is indeed the greatest form of Beauty.'

Elaborate ornaments whose designs are based on geometry constitute the basis of Islamic art, liberating artists from religious constraints. Further developing Arabic art and science, Westerners in the Renaissance introduced various symbols – letters for constants and variables, plus and minus signs, the × of multiplication – which made mathematics an almost universal language.

Ancient decorative schemes did not easily reveal their secrets even though astronomers such as Kepler attempted to unravel them. Only in the nineteenth century did we begin to understand their foundations. Mathematicians, often passionate musicians, such as Leonhard Euler (1707–83) or Joseph-Louis de Lagrange (1736–1813), made it their objective to

SEVENTEEN SYMMETRY SCHEMES, PETER STEVENS, 1981
Ancient cultures used all these patterns in decoration. Mathematicians were able to reconstruct them when advanced calculus used for crystallography was summoned to the task. Analysis then indicated that regular repetition of elements in a plane is restricted to these seventeen basic archetypes.
Handbook of Regular Patterns. *The MIT Press*

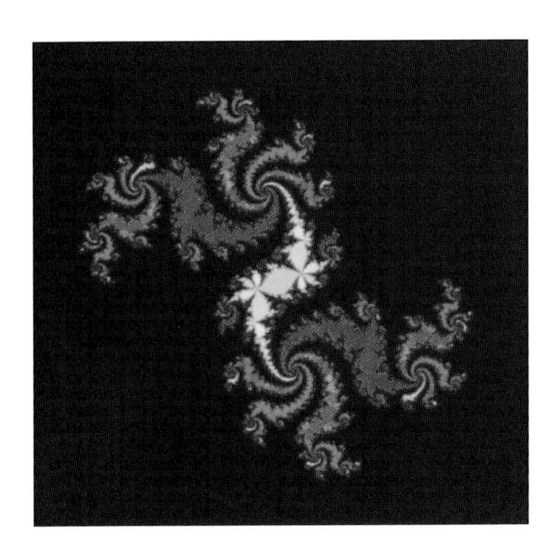

SELF-SQUARED *FRACTAL DRAGON*,
BENOIT MANDELBROT, 1982
When developed by computer, the formula $Z' = Z(Z-1)$ gives rise to a fractal image that looks like a dragon.

reformulate natural phenomena into a more handsome language than that of their predecessors.

Scientists are seduced by the aesthetic nature of theories when these are expressed with style. Ludwig Boltzmann (1844–1906) wrote: 'As a musician can recognize his Mozart, Beethoven or Schubert after hearing the first few bars, so a mathematician can recognize his Cauchy, Gauss, Jacobi, after the first few pages.'

Contemporary researchers believe that even the most intricate figures can be described in mathematical language by means of 'fractals' which are known to the public for their visual beauty. (A fractal image is made up of smaller identical structures, each of which represents the whole picture.) Of this computer-made art, its inventor, the mathematician Mandelbrot, thinks that: 'it is premature to ask if it will ever be able to compete with painting or photography, but it is legitimate and useful to raise the question.'

Artists and scientists aim at representing nature's structures, considered beautiful for their forms and the logic they embody, to create a coherent language for the arts, the sciences and nature's architecture, alike.

The German philosopher Johann von Herder (1744–1803), Goethe's friend and mentor, wrote: 'From stones to crystals, from crystals to metals, from these to plants, from plants to brutes, and from brutes to man, we have seen the form of organization ascend, and with this the powers and propensities of the creatures have become more various, till at length they have all united in the human frame.'

DETAIL OF THE AIDS VIRUS MODEL
Decorative patterns resemble natural shapes; most forms in nature, however, possess a sophistication that defies copying by scientists or artists.
J. -C. Chermann © Photothèque INSERM

SELF-PORTRAIT FOSTERS POND,
ARNO RAFAEL MINKKINEN, 1989
Nature's patterns, such as proteins, brains or bodies, are fundamentally asymmetrical. The physicist Wolfgang Pauli called God 'a weak left-hander', believing that asymmetry may be necessary for the universe to hold together.

From weaving to computers

Thousands of years ago, plants such as cotton were already cultivated for fabric production; elaborate patterns were woven according to ancestral techniques. There has been little change since; in fact, the design of ancient Peruvian textiles, for example, was finer than that of modern ones. Textile motifs are considered as one of the most ancient forms of visual and even of op art.

Much earlier, man's use of bone needles already marked an important step in the elaboration of clothing. Fabric remained a luxury in the West until the end of the Middle Ages, when the horizontal loom, which speeded-up weaving enormously, and the spinning wheel were invented. Sartorial fashion gained further ground in the fourteenth century with the introduction of knitting and the button.

The cloth business became the basis of Renaissance prosperity and allowed the Fugger and the Medici dynasties to establish multinational enterprises. These prompted the development of the double-accounting system, still used today. By setting in place an annual balance of profits and losses, the income of the business could be visualized and controlled. Luca Pacioli (1445–1540), who developed this system, was a protégé of the painter Piero della Francesca (1416–92); Leonardo himself illustrated one of Luca's books.

The Mathematician, Jacopo de Barbari, c. 1470
Barbari, also a mathematician, wrote a book on the Golden Section. This painting, in which a polyhedron seems to 'float', represents Pacioli practising geometry with the right hand, and arithmetic with the left.

Museo e Gallerie Nazionali di Capodimonte, Naples. Alinari-Giraudon

During the fifteenth century, businessmen carrying out large transactions found themselves drowning in complicated fractions (such as 5432434/2980764!). Simon Stevin (1548–1620), a mathematician from Bruges, thus developed a system to calculate decimal fractions. He published it in a 36-page booklet which immediately became famous. A polymath engineer, specializing in weight and length measurements, Stevin also developed musical theory.

Meanwhile, luxury silk, an even more lucrative trade than cloth, penetrated the West. According to legend, the 14-year-old wife of the emperor who built the Great Wall of China, discovered silk by dropping a cocoon into warm water, unwinding a glistening thread.

Bombyx (silkworm) larvae were transported from the East via Constantinople to Sicily and from there to northern Italy. The precious silkworms created a major problem for the French economy, as gold to pay for silk poured out to Italy. King Louis XI (1423–83) sought to stem the flow by converting Lyons from a local weaving centre into the European silk capital; a gigantic activity developed with fairs at international crossroads.

In 1741, around the time of the Industrial Revolution, a mechanical silk loom was invented by Jacques de Vaucanson, who made all kinds of automates. This was later improved by Joseph-Marie Jacquard (1752–1834), who introduced a system of punched cards that directed machines to weave elaborate designs – each card carrying a separate section of pattern. A similar technique is in use in today's power looms.

Research by Ada of Lovelace, a brilliant mathematician who invented computer programming, contributed to the workings of Charles Babbage's calculating machines – among them steam-driven 'analytical engines', which found no immediate practical applications. (A famous portrait of Charles Babbage was produced by using 24,000 punched cards.)

Statistical studies and logarithms all required machines, yet nothing much happened on the computer front for more than a century. Then the need grew for powerful calculating tools; during the Second World War the 'electronic brain' prototype proved its worth. Much work still remained to be done to turn this slow-working machine into a commercially appealing product . . . and few believed it would ever happen.

DETAIL OF THE AUTOMATED SILK LOOM,
JOSEPH-MARIE JACQUARD, 1804
The automation of silk manufacturing caused social upheaval, much like computerization does today. Moreover, the punched-card system that governed the machinery is essentially the same as the one employed in computers, two centuries later.

Ada Augusta King, Countess of Lovelace, c. 1840
The countess, daughter of the poet Byron, invented computer programming. One of the modern information languages bears her first name, ADA. Of the analytical machine she said: 'It weaves algebra patterns like Jacquard's silk loom weaves flowers and leaves.'
Science Museum, London. (Source)/Science & Society Picture Library

Early electronic calculating devices required thousands of vacuum tubes to transmit electronic signals. (The first computer – designed to crack German secret codes – was named Colossus because of its size.) Tiny devices called transistors were able to amplify electric signals and thus replaced the vacuum tubes. Since then, transistors have been packed on crystals made of fingernail-sized 'chips' of silicon.

Like many scientific inventions that develop in fits and starts over great distances, the development of computers showed that science, like art, does not always evolve in a linear fashion. Technologies and concepts ignored for years are reharnessed and perfected when a real need is identified.

In the eighteenth century, the discovery of punched cards for silk weaving eventually led to the invention of the computer . . . some two hundred years later. The transition from Andean weaving to information systems entailed much trial and error by generations of artists, craftsmen and scientists.

The United States of America Census, 1890
Electrically driven punched-card machines were used by Hermann Hollerith for the first large-scale census. He extended their use for data processing in other areas and founded the company later known as International Business Machines (IBM).
Scientific American

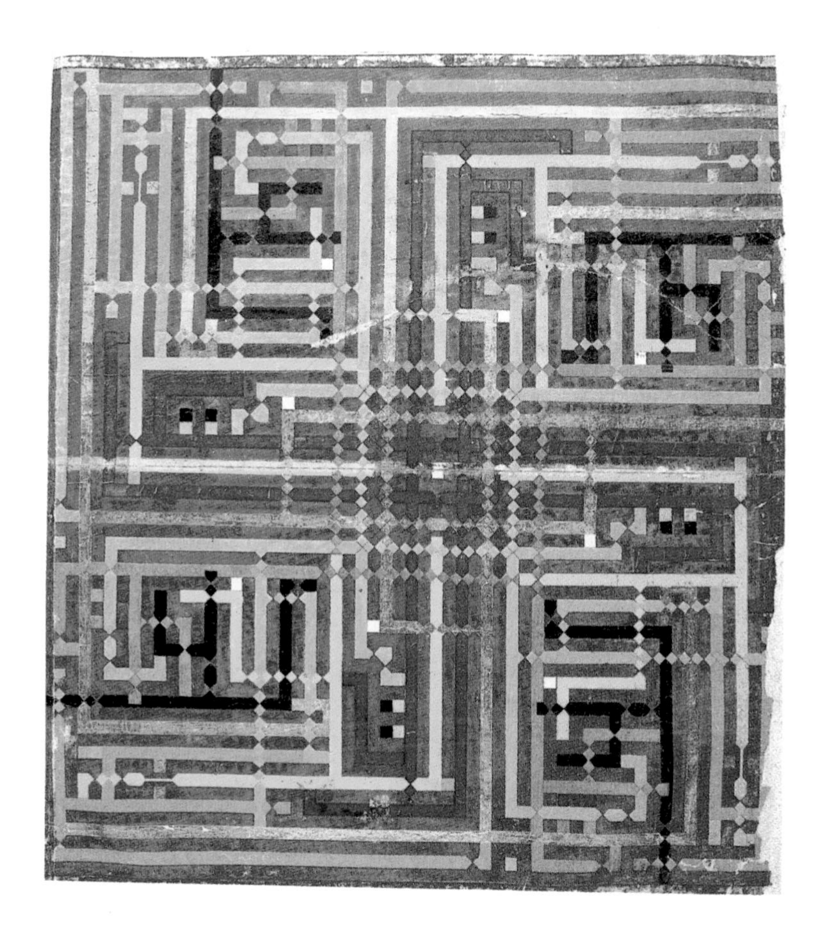

PERSIAN CALLIGRAPHY REPRESENTING THE NAME OF ALLAH, FIFTEENTH CENTURY
For Arabic calligraphers, 'the letter is the body and the number is the soul'. Ancient decorative shapes can look strikingly similar to microchip patterns – the latter were originally selected through design competitions.
H.2152. Topkapi Palace Museum, Istanbul

Pigments and materials

Man's fascination with patterns was paralleled by an early urge to experiment with colour. Dyes were made from substances of animal, mineral and vegetable origin. Introduced as early as 1000 B.C. by the Phoenicians in the Mediterranean Basin, fabric dye extracted from purple snails would be reserved for high-ranking citizens in Rome.

Deep blue pigment from lapis lazuli, a stone found in Afghanistan, came all the way to the monasteries of Ireland. Indigo came from India to Britain; pastel blue from Provence. Dyes were the basis of a luxury trade prompting chemists, botanists and mineralogists to seek samples from all over the world, for new colours and the technical means of obtaining them. Gradually, scarlet, blue of Saxony, Prussian blue and many others enriched the costly repertoire.

The first artificial dyes were the result of the Industrial Revolution when coal tar, used to illuminate the gas street-lights of London (1792), yielded by-products such as benzene. These were available in large quantities, waiting for scientists to find potential uses. The accidental discovery in 1856 of mauveine by William Perkin, aged 18 (while he was trying to synthesize the colourless quinine – a structurally related compound), marked the beginning of a true industry of pigments.

When synthetic dye was presented at the World Exhibition in London, manufacturers were reluctant to invest. In France, the Empress Eugénie launched the 'mauve craze'; a large-scale industry expanded rapidly with several derived colours and shades of blue, violet and green.

The springboard of today's chemical empires was in place and its implications for the industrial arts would be immense. As the Italian poet, Dante (1265–1321), said: 'Art imitates nature as well as it can' . . . and so does science.

The revolution in pigments spread to textiles. Around the time that the Lyons silk industry was almost devastated by parasites, fabric research moved in yet another direction: artificial fibres. Although Louis Pasteur found the cure for silkworm disease, the impetus to discover a means of producing cheaper textiles remained strong.

Artificial silk was promptly developed and, soon afterwards, synthetic stockings appeared on the market. A breakthrough occurred when Wallace Hume Carothers, working on benzene derivatives, patented nylon (1937) – thereby paving the way for a generation of fibres that could cover people, cars, aircraft, roads and buildings. These fabrics revolutionized the material and aesthetic worlds, and the lives of everyone they touched.

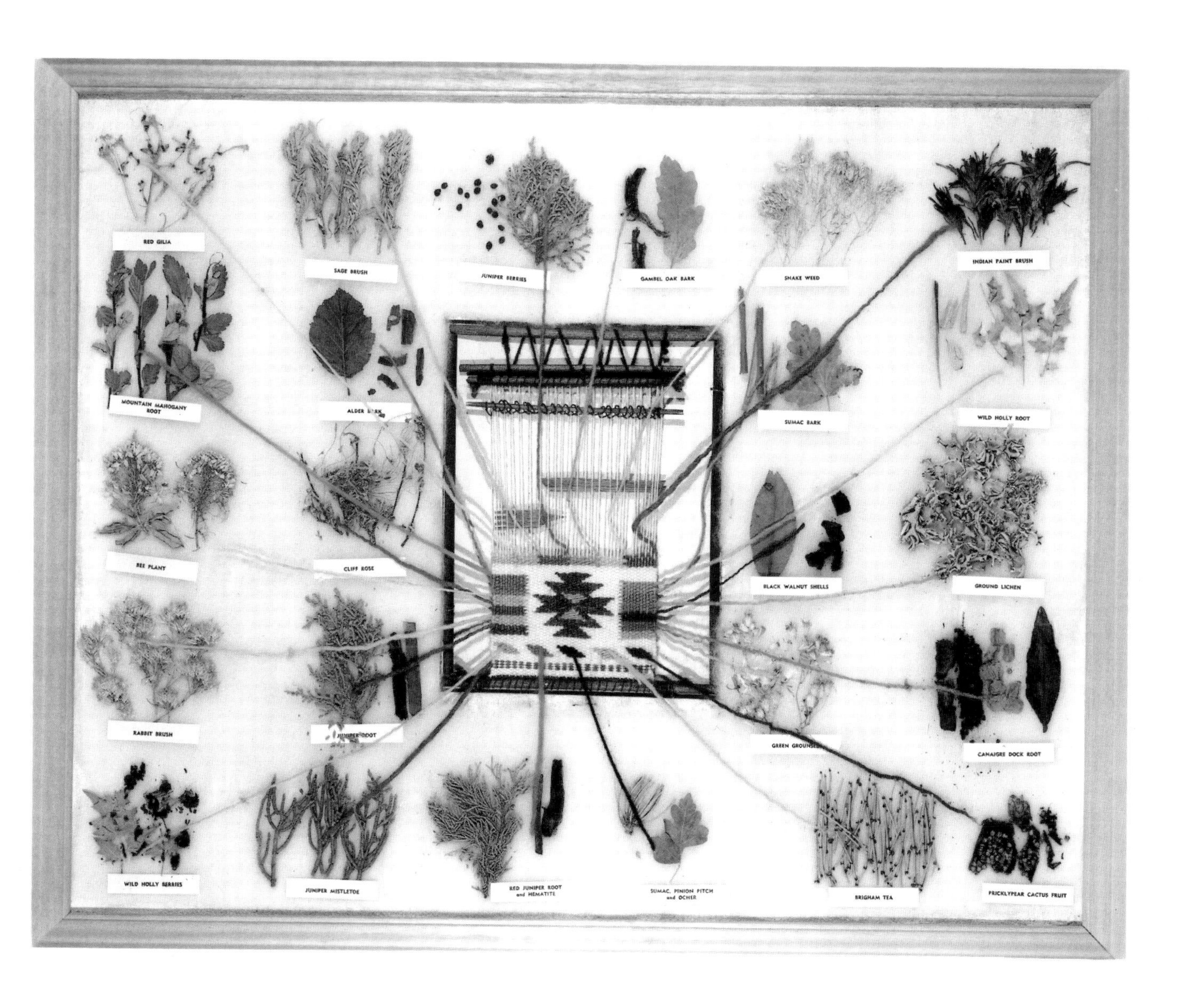

INDIAN DYE CHART MADE OF VARIOUS PLANT MATERIALS, NORTH AMERICAN INDIAN NAVAJO TRIBE, C. 1973–74
Ancient Peruvians produced almost two hundred hues. Carmine was extracted from a cochineal living in a Mexican cactus; ember red was imported from Brazil.

Approx. 55.2 × 70.5 cm. © The Cleveland Museum of Art, 1998. The Harold T. Clark Educational Extension Fund, 1974. 1060

Glass-blowers and telecommunications

Experimenting with earthen materials resulted in glass-making in the Orient 5,000 years ago. The Roman historian, Pliny the Elder (23–79), relates that glass was discovered by accident. According to legend, the momentous event occurred when a Phoenician crew landed on a beach and made a fire for cooking: all the basic ingredients happened to be there to produce the noble material.

However, the complex chemical process of glass-making requires a mixture of sand, ash and lime, combined with other substances (such as an alkali which lowers the vitrification temperature) and heated into a molten state. After quick cooling, the surface becomes covered with a transparent vitreous layer that can be worked in various ways.

This extraordinary material, compounded from ordinary ingredients, has inspired artists and scientists through the ages, who have penetrated its secrets and coaxed it to simulate the appearance of precious stones and metals. Some of the oldest known glass objects are translucent shards used for inlay work and coloured beads.

As knowledge and technical skills increased, molten glass was wrapped around a heat-resistant core made of lime-like material to which glass does not adhere.

The nucleus could be encased in several layers of glass with various decorative shapes and then destroyed, leaving the object intact.

COMPOSITE GLASS NECKLACE, EGYPT, C.1400 B.C.
© 1978 The Corning Museum of Glass

THRONE OF TUTANKHAMEN, C. 1350 B.C.
This throne was inlaid with precious and semi-precious materials: glazed ceramic, gold, silver and coloured glass paste.
Egyptian Museum in Cairo

The Greeks did not develop much interest in glass-making. The technique of blowing was invented somewhere along the Syro-Palestinian coast around the first century B.C., by then Roman territory. Blowing into a long hollow rod, craftsmen would force molten glass into a mould. Free-form blowing followed, allowing for diversity of form and greater translucency.

The Romans made great strides in glass-making. They invented ways to eliminate defective colour due to impurities, and developed techniques for applying decorative motifs, incising, engraving and gilding. Magnificent objects (oil lamps and perfume bottles), as well as sturdy glassware for everyday use, resulted. Production of the latter became a major industry.

GLASS 'CAGE CUP', ROMAN, C. 300 A.D.
This cut-glass bowl seems to be resting in a cage. The iridescence frequently seen in ancient Roman ware is a sign of glass decay.
Gift, funds from Arthur Rubloff Residuary Trust.

In the Middle Ages, glass was placed alongside precious gems in jewellery and reliquaries, while alchemists searched for the 'deep red' formula. It was believed that drinking from a ruby-coloured glass protected against many ills. Glass features prominently in legends, and toasts are still proclaimed when glasses are raised.

Chemicals, in the form of metallic oxides added during the melting process, greatly extended the colour range. The fruits of these experiments were, of course, the stained-glass windows of Gothic cathedrals in the North, and the sumptuous mosaics in the South. Apart from colour, however, most of the hard-won knowledge of glass-making was eventually lost to Europe.

In the Arab world, techniques continued to improve and reached an artistic height. The versatility of Islamic glass-makers resulted in some objects that looked as solid as rock crystal, and others that were translucent and paper-thin. One lamp bears the inscription: 'God's light is as a niche in which there is a lamp of glass, as if it were a glittering star.'

Venice then took over as the world's glass-making capital. Venetians developed mirrors and *cristallo* (crystal was invented by the Englishman George Ravenscroft) – a luminous material resembling rock crystal that made a clear resonant bell-like sound when struck. Venetians refined techniques used to produce *millefiori,* filigree, diamond engraving and enamelling.

Glass-making was restricted to the island of Murano, ostensibly to avoid the danger of fire in the crowded city, but also to prevent the glass-makers (who were forbidden to leave the island) from divulging precious professional secrets. Despite the law, many established themselves in Bohemia and in other royal courts. The Venetian style dominated Europe for nearly two centuries, until regional styles took root and began to undermine its pre-eminence.

Influenced by the Arts and Crafts movement, nineteenth-century glass artists such as the Frenchman Émile Gallé (1846–1904) and the American Louis Comfort Tiffany (1848–1933), developed an immense range of methods and materials and created objects of great beauty. Many of their stylistic innova-

MICROSCOPE OF THE DUKE OF CHAULNES, EIGHTEENTH CENTURY

European courts were fascinated by scientific toys based on glass; microscopes were often richly decorated. A Dutch ambassador to King Louis XIV described one of them as: 'almost one-and-a-half feet long, made of gilt brass two inches in diameter, mounted on three dolphins of brass, on a base of a disk of ebony.'

tions were based on rediscoveries of earlier Roman and Venetian glass-making methods, and of Egyptian and Japanese styles.

Innovative achievements today are due to the artist's autonomy following the invention of the small furnace. This piece of equipment made it possible for glass-makers to establish private studios and thus function as independent artists. It was technology that finally put them on a par with painters and sculptors.

CRYSTAL VASE, ÉMILE GALLÉ, 1892. DETAIL (ABOVE)
Gallé admired scientists and made this vase for Pasteur's jubilee. A microscope and a dog with symptoms of rabies are featured on the vase, inscribed with these lines by Victor Hugo: 'I move about in meditation, and at all moments a certain instinct obliges me to seek and find what lies behind the suffering of men.'
Institut Pasteur

The inherent properties of glass made it an obvious choice as an art medium, but scientists were as intrigued by the material as artists were. Convex lenses were known as early as 300 B.C. Much later, Arab scientists developed parabolic mirrors resembling those used in telescopes, and explained how lenses work.

SCHOLAR WEARING EYEGLASSES,
TOMMASO DA MODENA OR DA BARASINO, C. 1352
Scientists knew about lenses long before this first image of spectacles, invented in 1290, was painted
Seminario Treviso.
Photo Produzione Image & Communication SNC

The English philosopher, Roger Bacon (*c.* 1214–94) explored an extraordinary breadth of disciplines. Of optics, his favourite subject, he wrote: 'It is possible that some other science may be more useful, but no other has so much sweetness and beauty of utility.' For him, optics was almost a form of visual art. He studied the laws of reflection and refraction, and gave a theory of the rainbow as an example of elegant inductive reasoning.

Bacon made a description of lenses and eyeglasses, and expounded on their merits for the far-sighted. Lenses were then ignored by scholars for a very long time: three hundred years elapsed between the first recorded use of eyeglasses and the next major discovery dependent on glass, the telescope. Moreover, at the time no one guessed that it would be instrumental in revolutionizing man's understanding of the universe; the telescope's initial purpose was to serve the development of a new generation of artillery.

As the developing prosperity of the Western world brought about a growing need for private housing, plate glass was much in demand. In England, where window production was particularly intense, glass-making precipitated a timber shortage. Forests receded further and further from the factories, and it became obvious that an alternative form of fuel would have to be found.

VIEW OF THE GLASS CENTREPIECE AT THE CRYSTAL PALACE, LONDON, 1851

Sponsored by the Royal Family, the first World Exhibition was held in this gigantic glass greenhouse made entirely of prefabricated elements, which covered an area of 20 acres. The show was dedicated to the applications of art and science from nearly ninety countries. The latest agricultural tools, dinosaur bones, hi-tech musical instruments, etc., drew enormous crowds. The centrepiece of this giant fair, 'the gem of the transept', was a huge glass fountain.

Collection of Juliette K. and Leonard S. Bakow. Research Library of The Corning Museum of Glass

Glass production thus played a significant role in the Industrial Revolution. It was also the first industry established on American soil. Benjamin Franklin used glass equipment in his experiments with electricity. Modern chemistry and physics all advanced substantially with the help of glass instruments.

One of the most important of all inventions was, of course, the electric light bulb. In buildings, windows eventually became wall-size, resulting in the virtual elimination of visual barriers between indoors and out. Glass has changed our perception of light and our concepts of art and architecture.

It continues to change our lives. Analysis of the properties of this astonishing material has led to the creation of semiconductors used in electronic components. New forms, such as glass fibres invented by Narinder Kapary (1955), can transmit numerous high-resolution signals.

The first picture taken *in utero* by a camera placed at the end of an endoscope (a tube lined with glass fibres, inserted into the body), exemplified how glass contributes to the practice of medicine. Its role is spectacular in today's aerospace industry, and the telecommunications revolution in progress rests . . . on a glass foundation.

Throughout history, materials resembling glass – lacquer, enamel and amber – were considered to be imbued with special powers and integrated into crafted objects with important functions.

According to Joseph Needham (1900–95) – a prolific British biochemist and historian who spent decades writing a monumental history of science and technology in China, lacquer was a symbol of immortality, since 'it was present at birth as a crib, accompanied life as a spoon, then a chair, and could serve as . . . a coffin'.

Needham described lacquer as the first 'plastic'. This resinous substance, obtained from lac tree-sap, remains liquid if mixed with crab extract, which prevents hardening. Completing a single object could take up to several months, since twenty or more layers are required. The illusion of depth caused by the reflection of light depends on the build-up of layers and their optical quality.

Lacquer production thus demands intuitive knowledge of physics and high-level skills. The technique, which was also used to improve durability of ordinary tools, is said to have been known 3,000 years ago in China; it saw a marked growth when production became organized under imperial control, and it was further refined in Japan.

IMPERIAL LACQUER THRONE, CHINA, EIGHTEENTH CENTURY
First, a wooden core was fashioned in the shape of the final object. After meticulous smoothing it was covered with several layers of cloth impregnated with a mixture of liquid lacquer and clay. Each layer was carefully dried and polished before the next one was added until a final coloured layer, which could vary from bright to deep red, or black, was applied. The surface decoration was then worked out with the addition of gold, silver, shells and precious stones.

Enamel, originally used in finishing ceramic surfaces, became an artistic medium in its own right, with the technique known as cloisonné. This class of decorative objects was made by pouring liquid enamel into connected cells formed by metal wire and arranged in a variety of shapes and patterns.

The complex process which required the skills of the metallurgist, the coppersmith, the painter and the engraver probably originated in central Asia.

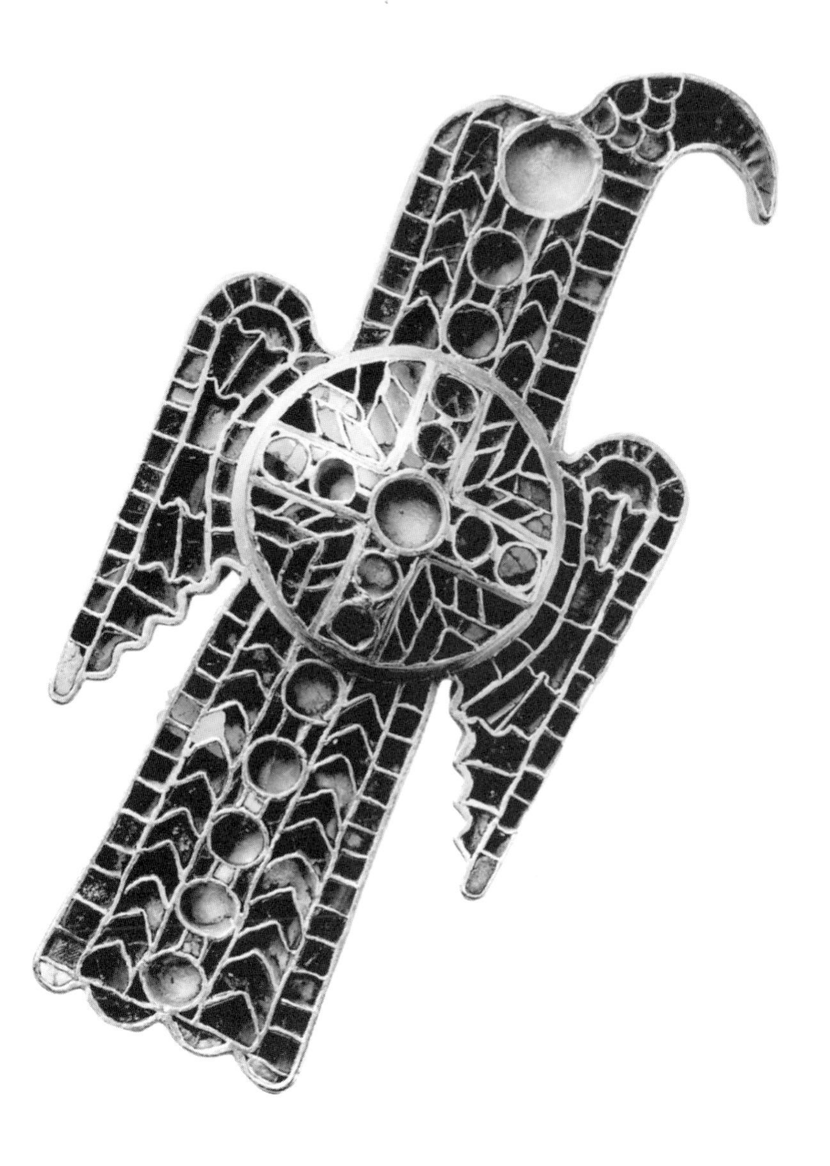

Cloisonné brooch, Ostrogoth, c. 500 A.D.
Western cloisonné may have been intended to simulate the look of Byzantine mosaics. It could frame enamel, gems or glass paste, and was frequently used by the Franks and the Goths, whose small transportable objects probably contributed to the development of stained-glass windows.
Germanisches Nationalmuseum, Nuremberg

Amber, another glass-like mythical material, is a fossil resin whose colour varies from pale yellow to deep red. Used in jewellery and ornaments, it was transported by the ancient trade routes from the Baltic, all the way to the Black Sea. In an Assyrian inscription, a king said that his caravans 'went fishing for yellow saffron in the sea under the polar star'.

The Ancient Greeks knew that amber, *elektrike,* had static electrical properties. Scientists have recently cloned the DNA of primitive insects trapped in this natural resin, hoping to bring extinct species back to life.

The potter's alchemy

Long before fashioning threads into garments and smelting glass, men moulded earth. In ancient finds, figurines tend to predate pots: magical intentions may have preceded practical ones.

Vestiges of ancient ceramics have been found in nearly every region of the world. Clay hardens naturally in high heat and perhaps, thousands of years ago, someone dropped a clay object in the fire and discovered that it hardened and retained its hardness ever after.

Experiments with progressively higher temperatures ensued and minerals were added, resulting in pots that were not only increasingly sturdy and useful, but also colourful and pleasing to the eye. Ceramics evolved into a most elegant art.

The Mesopotamians and Egyptians, using simple kilns, attempted to simulate in their potteries the look of precious marble and alabaster. Coating glazes that contained chemicals such as lead were admired for their lustre, moreover they reduced deformation of the object during cooling. These research efforts were probably related, technically, to the

RESTORED ISHTAR GATE, BABYLON, SIXTH CENTURY B.C.

The gates of Ishtar, goddess of fertility, were paved with brightly-coloured ceramic protective tiles. The Mesopotamians added copper powder to their clay to produce pottery the color of lapis lazuli.

Vorderasiatisches Museum. Staatlichen Museen, Berlin. © bpk. Photo Klaus Göken, 1992

manufacturing of glass. The practice of all these crafts carried symbolic meanings. Meanwhile, the introduction of the potter's wheel, in Mesopotamia around 3500 B.C., made regular forms and smooth surfaces relatively easy to achieve.

In Greece, the potter's wheel became widespread around 1000 B.C. The ingredients used were very basic, even though skills were amazingly sophisticated. Artistic heights were reached around 500 B.C., when red figures replaced black ones, allowing for livelier depiction. (The entire vase turns black when the inlet of the kiln is closed; when the air vents are opened again, the iron in the clay body returns to its red colour.)

Through experimental cycles called reduction and oxidation, using layers of clay-slip with differing composition, Greek artists suggested ideas to the natural philosophers that became the premises of geology and chemistry. But, above all, Greek potters transformed a simple craft into one of the most admired art forms, which more than any other has brought antiquity to life.

DETAIL OF A GREEK VASE, FIFTH CENTURY B.C.
The early figures on Greek vases featured black on a red background. Details were achieved by scratching the figures; these could never be very expressive and were thus often presented in profile. Greek potters then switched from black to red images, which allowed for more detail, painted with a fine brush on the figure.
Staatliche Museen zu Berlin © bpk. Photo J. Laurentius

FIGURINES OF CAMELS, GLAZED POTTERY,
TANG DYNASTY, CHINA, 618–907 A.D.
The Chinese experimented with a wide array of pigments to expand the colour range of their ceramics.
Objects 466 & 465 neg. #T4-217c.2.
University of Pennsylvania Museum, Philadelphia

In China, pottery production developed much earlier than in the West, under the Shang dynasty (1766–1122 B.C.) – the same period that saw the appearance of famous cast-bronze vessels. The two techniques were closely related: pottery vitrifies in a temperature range similar to that necessary for bronze smelting.

A long tradition of firing at ever-higher temperatures gave Chinese stoneware a harder and more fused body than Western ceramics. True porcelain was made by firing above 1,000 °C China clay (kaolin) mixed with *petuntse,* a combination which vitrifies at low temperatures and increases translucency. This technique, already known in the sixth century, was not fully exploited until later.

The Chinese identified porcelain by the clear sound it emits when struck, although the term was reserved for white, translucent ware: white was the symbol of purity. Chinese potters also liked to give their creations the appearance of jade, which they considered the noblest material. The typical green 'celadon' was obtained by firing iron contained in the glaze. (The opalescent quality is due to the reflection of scattered light by water vapour bubbles trapped during the firing process.)

Increasingly subtle shades of yellow, amber and brown were attained as chemical experimentation progressed, and pigments continued to be traded across Asia. For example, adding Persian cobalt under particularly exacting conditions resulted in the typical Ming blue. Aubergine, green, pink and yellow enamels produced families of Chinaware that were exported worldwide, and state-owned factories developed all over the country.

By the end of the sixteenth century, many clay deposits were depleted and the famous tradition declined. The first sample of China clay was probably brought to Europe by Portuguese navigators around 1520. Westerners hoped to make porcelain instantly, but it took them several centuries to succeed.

Islamic pottery made its way north from Spain and Italy. The Arabs' knowledge of earthenware production was fragmentary because the basic Chinese ingredients were lacking. Pottery was fired at low temperature, shape was lost in the process and the objects remained relatively soft. Nevertheless, Arab techniques flourished in Europe, where brightly coloured majolica and faenza – after the names of the manufacturing centres – became popular export products.

The Renaissance saw advances in the natural sciences with the emergence of artist-scientists like Frenchman Bernard

Palissy (1510–89). He was a maker of stained-glass windows who developed pottery glazes, and rediscovered enamelling techniques. He passionately researched a wide range of geological and chemical issues (and went as far as burning his furniture to keep his furnace going). An expert on natural history, especially fossils, Palissy was particularly good at representing naturalistic subjects, widely imitated in the following centuries.

OVAL DISH, AFTER BERNARD PALISSY, SIXTEENTH CENTURY
During the Renaissance, artists showed interest in the earth sciences. Palissy also designed an artificial grotto for Catherine of Medici's garden.

Lead-enamelled earthenware. # 60.8 Arthur Mason Knapp Fund and Anonymous Gift. Courtesy Museum of Fine Arts, Boston.

Meanwhile, the race was still on to duplicate the quality of Chinese pottery. Around 1708, Johann Böttger, experimenting with very high temperatures, finally uncovered the secrets of hard porcelain. His effort was funded in large part by the Elector of Saxony, who actually expected that it would yield . . . gold!

The secrets of porcelain-making were soon leaked and factories sprang up throughout Europe. The pottery at Sèvres in France (founded in 1756) was famous for its blue ware, and each court wanted to compete with its own distinctive colours and shapes. Royal feasts were ideal occasions for imagining new styles.

PHARMACY JARS, SIXTEENTH CENTURY
Brightly painted pharmaceutical jars were commonly produced in Italy, to contain a mixture of more than seventy ingredients called *Theriaca*, considered a cure for all ills until the nineteenth century.
Science Museum, London.
Photo 1970. Wellcome Institute Library

While artists created porcelain masterpieces, scientists such as the Frenchman, René de Réaumur (1683–1757), researched the chemical and physical rationale for the vitrification process of earth, for which pottery appeared to be a good model. Samples of various materials were sent to him by explorers from the far corners of the world.

Research was also being conducted in the United Kingdom. Several techniques for making porcelain had been tried before the chemist Josiah Wedgwood (1730–95) made his breakthrough. Having a strong interest in antiques, he produced an exceptional stoneware called 'jasper', which could be stained lavender or green, and had the appearance of antique translucent cameos.

In order to carry out his experiments with clay, he invented a special instrument for measuring high temperatures. Wedgwood founded a modern ceramics production facility which boasted the first factory to use steam power. He developed the basis of a fully integrated industry – starting from research and development to production and marketing.

Among Wedgwood's acquaintances were Erasmus Darwin (1731–1802), naturalist and poet – Charles' grandfather, Joseph Priestley (1733–1804), the famous chemist who discovered oxygen, and James Watt (1736–1819) who developed the steam engine. Most of them were distinguished members of the Lunar Society promoting the arts and sciences. They were called 'lunatics' because monthly meetings took place on the Monday nearing the full moon, so that participants would have an easy time seeing their way home.

England's scientific community was flourishing. Increased prospection of coal and ores stimulated interest in fossils and rocks: geology became an important science because the building of canals and railways depended on it. Geology became so popular that Harriett Martineau (1802–76), a political and historical writer, commented: 'the general middle-class public purchased five copies of an expensive work on geology to one of the most popular novels of the time.'

A new theory of the earth's formation was then advanced, naming heat as the principal agent of the formation of the earth's crust. It contradicted the view that sedimentation and crystallization of water were principally responsible.

WEDGWOOD PYROMETER, UNSIGNED, *C.* 1780
Wedgwood invented a device which became a reference for measuring temperatures, particularly useful in pottery-making.
By permission of the Trustees of the National Museums of Scotland

JASPERWARE COPY OF THE *PORTLAND VASE*, JOSIAH WEDGWOOD *C.* 1790
Wedgwood symbolizes the exceptional relationship between science, industry and art that existed during his time. He had a particular taste for neo-classical art and founded a ceramics industry in Etruria (Staffordshire, UK), a name inspired by the antiques craze triggered by the discoveries at Pompeii. This vase is a high-quality copy of a then recently discovered Roman amphora.
© The Board of Trustees of the Victoria & Albert Museum. V&A Picture Library, London

Eventually, *The Principles of Geology* by Sir Charles Lyell (1797–1875) explained the changes as a result of an ongoing process rather than an isolated cataclysmic event, such as a flood or a volcanic eruption. Lyell's theory would greatly inspire Charles Darwin while travelling around the globe.

In the meantime, the small-scale pottery had become a large industry. Wedgwood's signature, cream-coloured earthenware which he was permitted to call 'Queen's ware', became a household staple, and his lavender ware can still be purchased all over the world.

Two hundred years ago, science and technology helped to turn an antique art form into a pleasing commodity. According to scientist Cyril Smith: 'it was, however, ceramics made by Neolithic creators who unwittingly caused the atoms in the clay to lock tightly together through the action of heat, that marked the true beginning of material sciences.'

FIGURE OF VOLTAIRE,
JOSIAH WEDGWOOD, C. 1775
Wedgwood immortalized Voltaire, whose companion, Émilie du Châtelet, was a knowledgeable mathematician. She translated Newton's notes into French.

By permission of The Potteries Museum, Stoke-on-Trent

The power of metal

Of all the earth's materials, metal is the most fascinating. In an oriental myth, the birth of metal is described as the result of the death of a superior being. Lead was extracted from his head, tin from his blood, silver from his marrow, copper from his bones, steel from his flesh . . . and gold from his soul.

Since stone knives were as sharp as blades, metal at first had no practical advantage and was probably used for magical purposes. Some metals occur naturally and are easily detected as shiny, malleable lumps on the earth's surface. The earliest metal objects, mostly small ornaments of gold and silver, were directly cut and carved.

According to the Greek historian Herodotus (*c.* 484–20 B.C.), Pactolus was the river with the richest source of gold in the world. This precious metal could be simply scratched from rocks and used as payment. King Croesus (sixth century B.C.), from the nearby region, introduced gold coinage. Meanwhile, several civilizations developed goldsmithing skills to a high degree.

Centre-bead and two Triratna-shaped necklace pendants, India, Sunga Period, 185–72 B.C.
Etruscans and Indians mastered filigree and granulation techniques exceptionally well. Minuscule spherical beads of pure gold were heated and rolled until they became solid, then thrown in water or dust before being chemically fixed to the ornamental piece.

Gold repoussé with granulation. Height 5.7 cm. © The Cleveland Museum of Art, 1998. John L. Severance Fund, 1973.66-.68

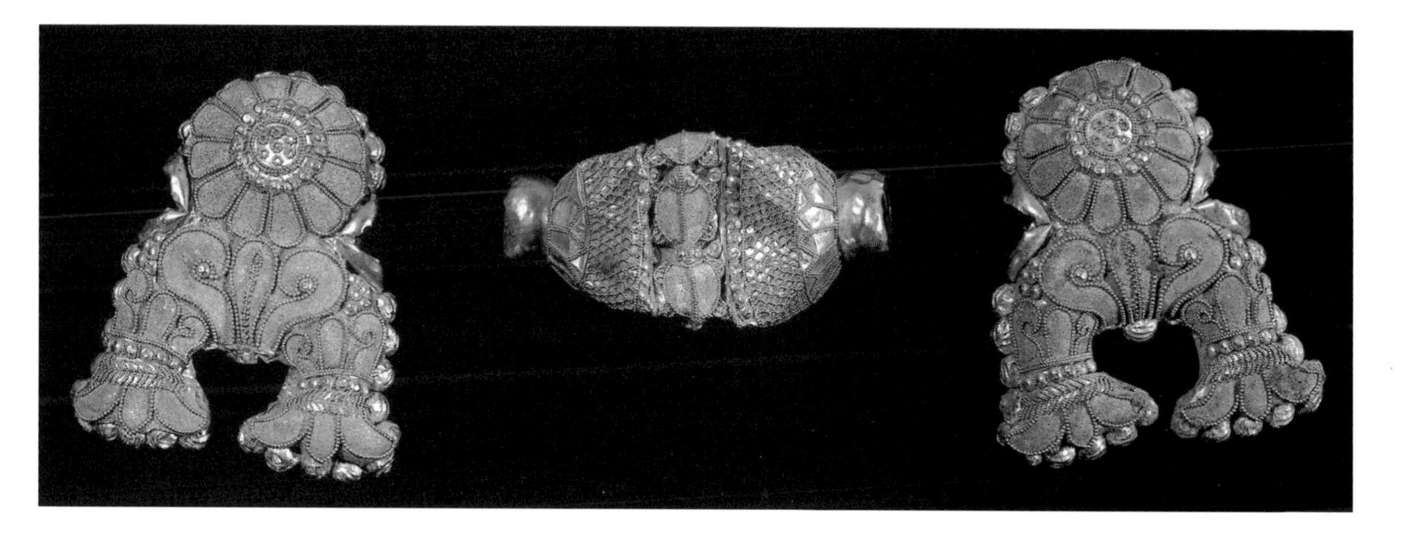

Unlike gold, most metals are not found in their pure form but have to be extracted and smelted. About five thousand years ago, it was discovered that, when subjected to fire, the green stone malachite and some other minerals produced red copper that could be hammered, cast, moulded, reheated and reshaped. The possibilities for making ornaments, vessels and tools increased dramatically as a result. But there was one major drawback: copper is soft.

The problem was overcome when it was, perhaps accidentally, discovered that when copper was combined with tin – which is even softer – an alloy was created that is surprisingly harder and more durable than either of its components. That tougher material is bronze (*c.* 3000 B.C.).

It took scientists until the twentieth century to understand the mechanism of this phenomenon, which is caused by changes in atomic structure. (Pure copper is like a wafer of parallel planes which slip easily when the metal is bent. Tin atoms act as anchorage points within the copper structure, preventing the planes from sliding.)

The amount of tin in bronze can range from 5 to 20 per cent; ancient Shang bronzes are usually found to contain the ideal proportions for creating the highest quality material. Many

ELEPHANT VASE WITH ELEPHANT COVER, BRONZE, SHANG, *C.* TWELFTH CENTURY B.C.

To make such an object, separate piece-moulds were impressed on clay models; the 'negatives' were carved and incised. When the pieces were assembled, molten metal was poured into the cavity between the core and the mould. In early vessels, bronze leaked between adjacent segments, producing 'fins'. In later times, these fins were integrated into the design of the finished objects. (Note the elephants' tails.)

 # 36.6 Courtesy of the Freer Gallery of Art, Smithsonian Institution, Washington, D.C.

ancient civilizations worked with bronze, but the Chinese attained the greatest mastery of the medium.

On the other hand, the Sumerians called iron the 'metal of heaven' because it was delivered by meteorites; Homer even put iron on a par with gold. In its pure form, it is also rather soft; man-made iron from terrestrial ores was developed late because processing required high temperatures. When people realized that it could be used for making hard tools and weapons, life was transformed.

Hence the existence of an early Stone Age, then a Bronze Age, followed by an Iron Age. (This categorization was drawn up by a Dane, Christian Jurgensen Thomsen (1788–1865) who, while searching for megaliths, found artefacts which he classified according to their constituent materials.) Thus materials and their production technologies form the basis of the chronology of civilization.

In the Far East, it was known by the fourth century B.C. that iron containing a small amount of carbon was easier to melt and cast than its purer counterpart. These experiments marked the first steps in the development of metallurgy. In order to attain high temperatures in the kiln, the Chinese covered its inside walls with heat-reflecting materials. In addition, large-scale use of furnace bellows driven by waterwheels contributed to the astonishing development of iron and steel in China, around 1000 A.D.

The Japanese, on the other hand, raised steel-working to an art that culminated in the production of ceremonial swords. Making one that was both flexible (having a forged iron core) and had hard cutting edges (made of steel) could require up to twenty phases of repeated heating, hammering and folding. The fabrication process was closer to a religious ritual than to a practical manufacturing procedure, however.

NYUDO KOTETSU DAGGER, TONTO, EDO PERIOD, JAPAN 1602–1867

The Japanese fashioned steel swords into functional works of art. The ornate scabbards were highly individual.

Various metals and lacquer. Overall length 41.1 cm, blade 28.5 cm. © The Cleveland Museum of Art, 1998. Gift of Bascom Little Estate, 1974.56

While Far-Eastern craftsmen created metal masterworks, Westerners were still struggling to fabricate rough cast iron. The Ancient Greeks performed a significant mental feat in hypothesizing that physical elements underlie the world's creations. By doing so, they stated that the universe could be ordered and framed. They had intuitive knowledge of the structure of materials: *atom* means indivisible in Greek and atoms in combination were named *moleculon*.

The belief that the four elements could blend into an infinite number of combinations to create phenomena, remained popular until well into the Renaissance. Alchemical concepts such as these flourished independently in Eastern cultures, too. In Europe they centred around the Philosopher's Stone, an imaginary object that would turn base metal into gold.

Mary the Jewess was credited with having invented the double-boiler and distillation. While studying the cycles of life and death, recording the secrets of metal, she said: 'Unite the male with the female and you will find what you seek.' Maturation of metal was indeed believed to accelerate the rhythm of nature. The alchemist Cleopatra (fourth century A.D.) wrote a book on the transmutation of gold.

In the eighth century, Jabir, known as the Mystical Arab, made important theoretical and experimental steps. Much later, a unique textbook on mining by Georgius Agricola (1494–1555) gave Western research a hefty impulse, but it was not until 1722 that a scientific treatise was written on the conversion of iron into steel. The combined knowledge of geology and metallurgy slowly paved the way for chemistry.

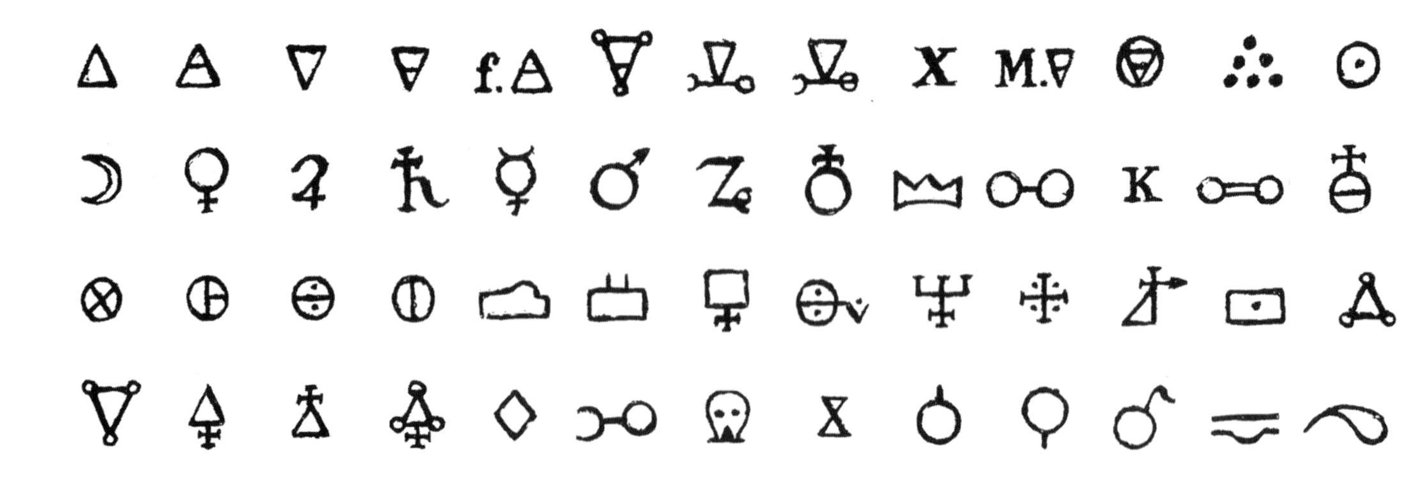

SYMBOLS USED IN ALCHEMY UNTIL THE EIGHTEENTH CENTURY
Chemistry was long a domain of fantasy; flowers of zinc, diaphoretic antimony, butter of antimony, vitriolated tartar, etc., were part of the poetic language.

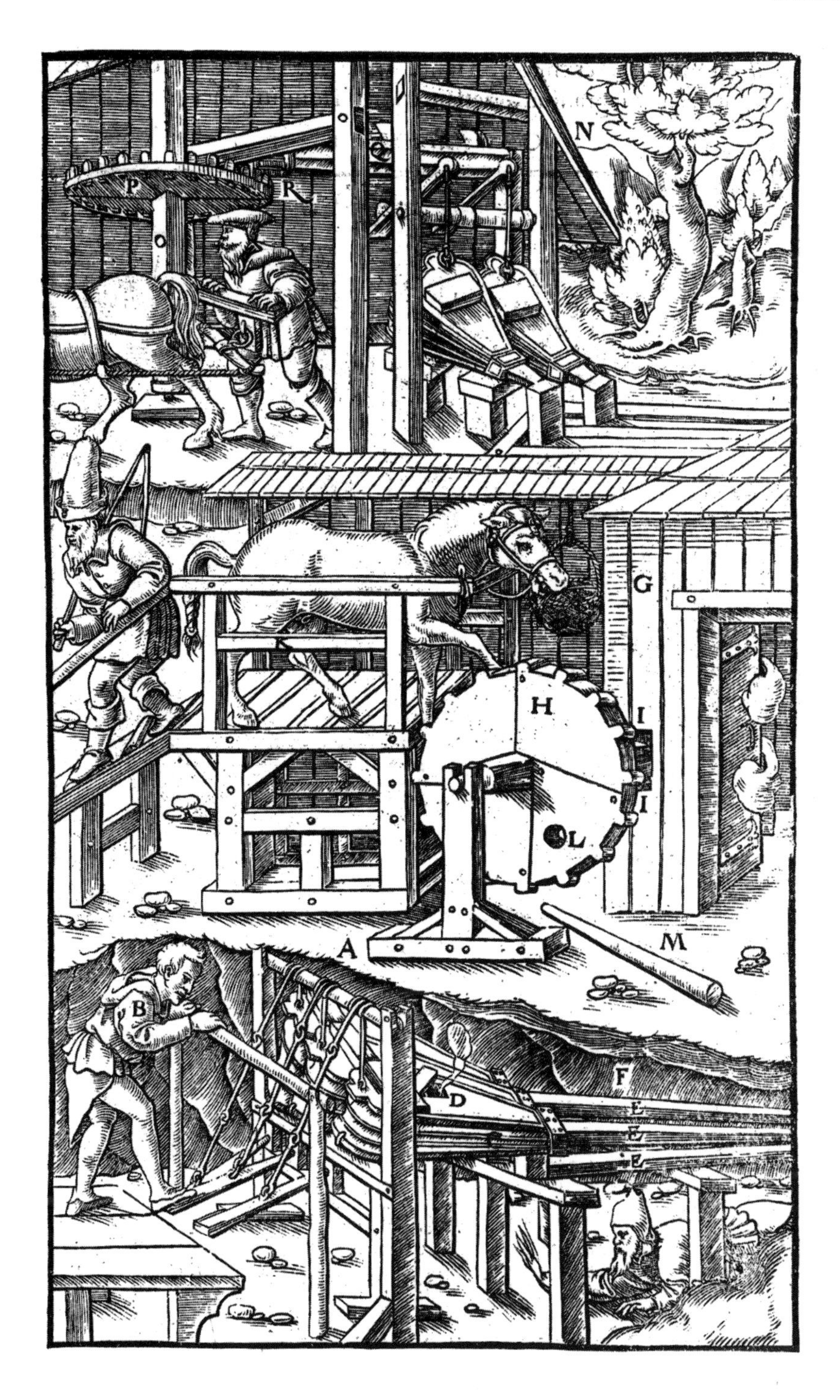

Mine Ventilation in *De Re Metallica*, Georgius Agricola, c. 1556
Agricola wrote a textbook on mining which served for two hundred years as the definitive work on the subject. He delayed its publication by decades in order to produce the 291 accompanying illustrations (and it was published posthumously). Agricola offered a detailed description of where mineral veins are found. He introduced the word 'fossil' (believing that it was an odd rock that looked like bones). The silver mines near Joachimsthal, where Agricola worked, gave their name to the thaler, which in turn became the dollar.
Science Museum, London. (Source) Science & Society Picture Library

SILVER-MINING COMMUNITY, FIFTEENTH CENTURY

Österreichische NationalBibliothek, Vienna. Photo Bildarchiv, ÖNB

The chemist's futuristic materials

During the eighteenth century, the study of combustion and respiration stimulated a particular interest for gas (pronounced 'chaos' by the Dutch scientists who coined it that way). The English consolidated the experimental knowledge of gases, while the French developed its theoretical properties. Antoine Laurent de Lavoisier (1743–94), drew the conclusion that air is made of oxygen and nitrogen; water was not an indivisible element either, as previously thought, but a compound made of oxygen and hydrogen.

A system was developed for classifying elements that could not be decomposed into smaller units. Each atom was considered to have a specific mass that remained unchanged during chemical processes. Laws that account for conservation of matter during chemical reactions could then be established. From there on, progress in understanding the structure and synthesis of matter was vertiginous.

A Russian visionary, Dmitri Mendeleyev (1834–1907), began to list the elements in order of their atomic masses. He described a basic table of 63 and correctly predicted the discovery of many others; today, their number exceeds 110. Although we know that the atoms, too, can be split, being made up of smaller particles, Mendeleyev's model is still applicable.

Progress in chemistry resulted in immediate applications in the artistic domain. A close friend of Count Alessandro Volta (1745–1827), the inventor of the battery, developed it to extract metals such as gold in order to plate sculptures and crafted objects. Depositing a coat of metal on to a core by means of electrolysis became instantly popular.

But the public demanded ever new products and the time was ripe for materials combining atoms into long chains called polymers, through elaborate chemical processes. An early commercial application of such products was as a replacement for expensive ivory for piano keys and billiard balls.

Alexander Parkes (1813–90) developed the first synthetic polymer, precursor of the gigantic plastics business. By-products of the coal and oil industry soon yielded a plethora of compounds (polystyrene, polyvinyl, neoprene, polyurethane, silicone, etc.) which reshaped decoration, painting, sculpture and architecture.

Thanks to computerized modelling techniques, and powerful microscopes which enable us to visualize atomic structures,

−
+
mA
V
V
reference electrode
anode
cathode (statue)
V = volt
mA = milliampere

GOLD-PLATING BY ELECTROLYSIS
Electrolysis was adopted on a large scale for reproducing works of art. Pre-Columbian Indians may have used a similar technique to gold-plate ritual objects.

matter can literally be created to measure. Atoms can be arranged in order to produce specific features such as conductivity, strength, and resistance to fire or bacteria.

Cyril Smith said: 'With the development of material sciences in the twentieth century, the scientist is now finding out something the artist has always known: imperfections in symmetry are the key to a higher level of interest. . . . And, of course artists did not think in terms of physicists' theories, but they did select the materials that would behave in the way they wanted.'

Since ancient times, decorators have inspired scientists by manipulating and alloying substances. Until the nineteenth century, nearly all available materials had been known for 5,000 years. Noble objects made of glass, porcelain and bronze are being progressively replaced by synthetic ones.

Although technology is overtaking a field traditionally dominated by artisans, together, artists and scientists keep fueling innovation in the decorative arts.

MOLECULE OF KEVLAR®
Long chains of atoms result in fibres that are several times stronger than steel, and yet relatively lighter than nylon. They can be used for bullet-proof jackets as well as a huge scenic decor. Seen through a microscope, this strand looks like a delicate necklace.
Courtesy of DuPont de Nemours

OPTICAL COMPUTER WAVE PRINT FABRIC

Synthetic materials have revolutionized clothing fashion by making it available to all.

W & L. T. spring/summer 1996. Photo Chris Moore

Painting and cognition

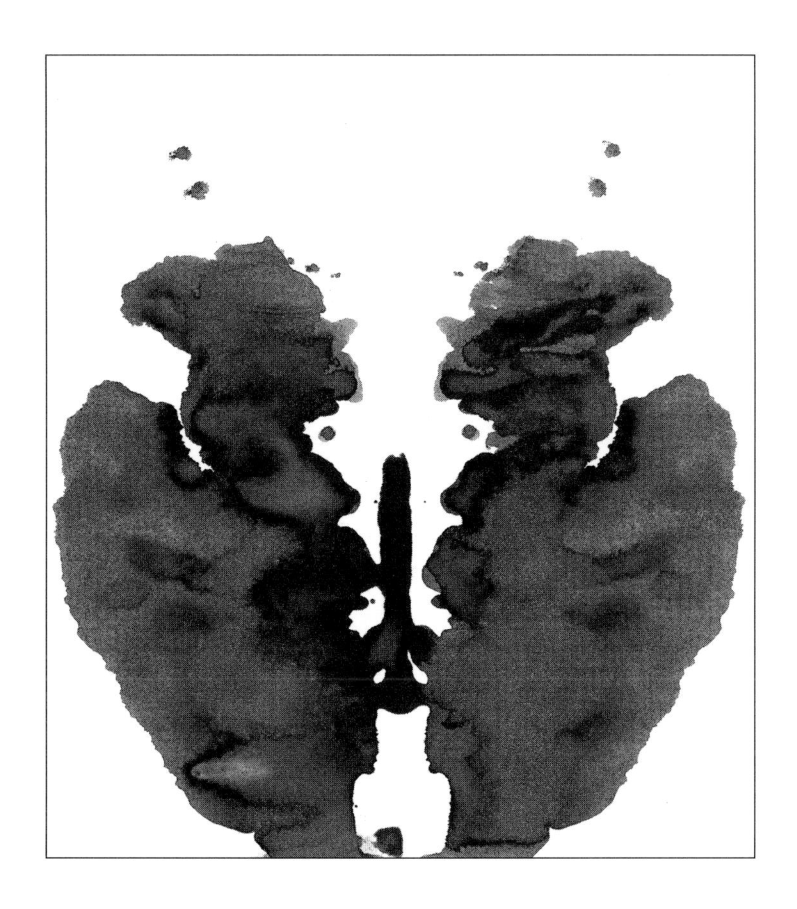

CHILD'S DRAWING
Rorschach, a Swiss psychiatrist who almost became an artist, introduced a psychological test, which is still used, based on patients' interpretations of abstract images. The artist's rival to the ink-blot is Alexander Cozens' system of blot drawing which was published a hundred years earlier, in order to stimulate artists' imagination.

Psychology in painting

Painters have always used colour to manipulate human emotions. Greek natural philosophers contended that all colours were made from varying amounts of four basic hues, as all things were composed of various proportions of four elements.

Theories based on combination of colours and elements – and comparisons with musical tones and planetary motions – were prevalent during the Renaissance, Leonardo developed such a theory. Feelings such as love and hate were believed to evolve cyclically in the same way as attraction and repulsion could dominate the universe.

The fundamentals of colour theory were, however, set forth in Newton's *Optics*. Later, colourism was extensively investigated by the chemist, Michel-Eugène Chevreul (1786–1889), the director of dyeing at the Manufacture des Gobelins, where a multitude of colours were made. Chevreul classified more than 15,000 hues and proposed scales and devices to codify them; he analysed the visual effects of colour juxtapositions.

Measurements allowed him to observe a number of intriguing effects. For example, red makes adjacent surfaces look greener, which led him to say: 'Where the eye sees at the same time two contiguous colours, they will appear as dissimilar as possible, both in their optical composition and in the height of their tone.'

In conjunction with Chevreul's theories, research by Eugène Delacroix (1798–1863) resulted in smoothing contours, in contrast to classical sharp lines. This work had a marked effect on the Impressionists who, although they often tended to oppose rationalism, were in fact the true harbingers of Newton's theory.

Following in their footsteps, Camille Pissaro (1830–1903) said: 'Neo-Impressionists should aim to seek a modern synthesis of methods based on science, on Chevreul's theory of colours and on the experiments of the physicist, Maxwell.'

Such debates fascinated the public and prompted painters to explore weight and mass by means of advancing and receding flat planes of colour, whose significance became discussed in terms of 'relativity'. Cézanne's facets, Matisse's patterns and Picasso's cubes have taken the viewer from extreme achromatism to splashing pigment. And the panel of media to produce colour has now been broadened to pixels.

NADAR INTERVIEWING MICHEL-EUGÈNE CHEVREUL, 1886
This photograph, taken by the son of Nadar (Félix Tournachon) was part of a series of interviews given by Chevreul, the leading colour specialist who was 100 years old at the time. His spoken words were recorded under each picture. Nadar had hoped to make an audio recording of these interviews, but the technical means for doing so were not yet ready.
Courtesy George Eastman House

The field of colour investigation has extended from eye physiology, to psychology and even to cognitive sciences aimed at understanding how the brain processes knowledge. While artists empirically discover the laws of colour, researchers create artful theories to describe the emotions they produce. In the course of this exercise, painters more than any other type of artist have often preceded scientists.

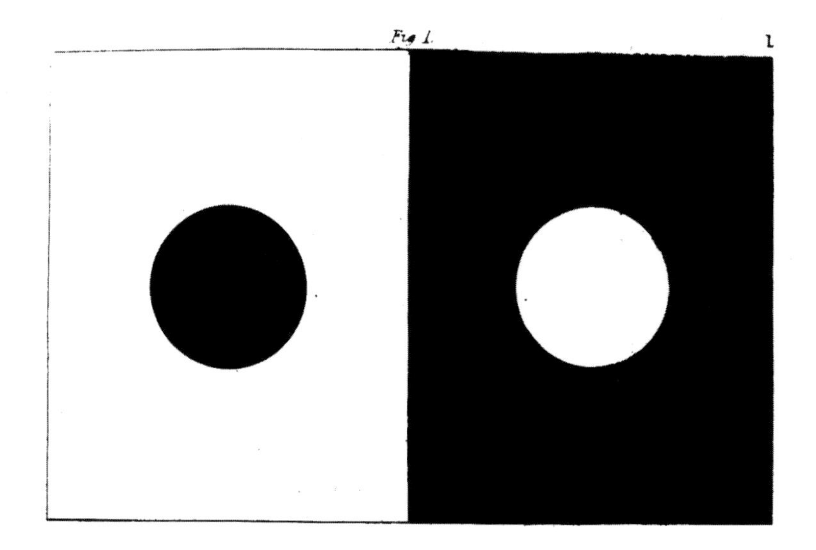

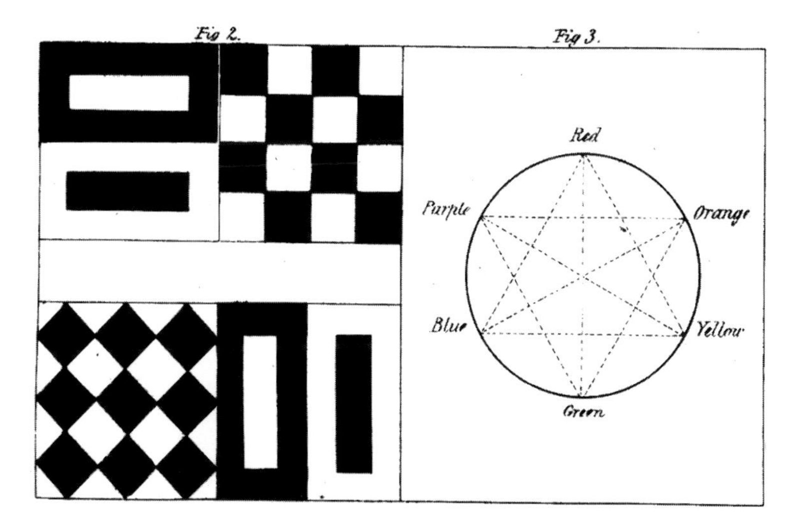

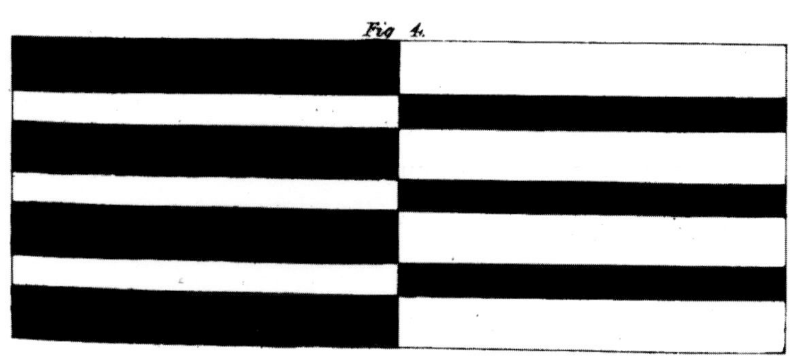

COLOUR WHEEL AND CONTRAST DIAGRAMS,
JOHANN WOLFGANG VON GOETHE, NINETEENTH CENTURY
Goethe contested Newton's views on colour and developed a psychology-oriented theory. Nineteenth-century scientists brought mechanical theory into correspondence with psychology. One century later, optical art developed by the Hungarian, Victor Vasarely – initially a medical student – suggested movement as a result of studies on psychology, chromatism and cybernetics.

Stiftung Weimarer Klassik. Herzogin Anna Amalia Bibliothek. Photo Sigrid Geske

Painting in two dimensions

Towards the end of the Stone Age, three forms of pictorial representation had developed: the imitative or realistic form, the abstract or decorative form, and the informative form. Colour was already exerting its power: the white, red and black triad was used in early funerary rites as well as in tribal masks.

Colour interpretation and perception is a culturally specific matter. For example, the Maoris identified over a hundred different types of red, South American Indians were sensitive to several dozens of varieties of green, and the Eskimos gave distinctive names to the different shades of white in their environment.

During the Stone Age, physical methods for grinding pigments had already been combined with knowledge of how to extract them from nature. Black and blue were produced from charcoal, ochre yielded yellow, red and brown. Pigments were mixed with animal fat to make paint that could be rolled into little sticks like crayons, or blown through hollow bones – or applied with a brush made of horsehair, feathers or weeds.

The cave paintings at Lascaux and Altamira are sufficient proof, if any were needed, that painting is not just a matter of materials and techniques. Artists must have been capable of great visual organization. Indeed, the perfect integration of their works into vast architectural underground spaces, 'Cathedrals of the Stone Age', show that they mastered the conceptual processes of the visual arts that are in use today.

AUSTRALIAN ABORIGINE PAINTING

Scientists have learned much about early times from clues left by artists. Naturalistic portrayals of animals, some now extinct, were created by the aborigines. 'X-ray type' painting indicates that this so-called 'primitive' society was interested in anatomy.

UNESCO/Thierry Joly

Frescoes (*fresco* means 'fresh' in Italian) on carefully man-made white-plaster walls, appeared as early as 6000 B.C. in the Middle East where, for example, enormous red bulls surrounded by tiny humans were powerfully portrayed.

Ancient Egyptian tomb paintings have survived, thanks to the arid desert environment, and still surprise us with their vivid colours. On plaster, paint dries almost immediately and the background needs constant humidifying; artists had to work fast, which perhaps added to the liveliness of their frescoes. The richness and variety of their colours are the products of a society familiar with the embalmer's art and the properties of a wide range of plants and pigments.

PAINT BOX OF VIZIER AMENEMOPET, EGYPT, EIGHTEENTH DYNASTY

Natural and artificial pigments are contained in this box; the depletion of the black cake suggests that it may have been used for writing, as well as for painting.

21.6 × 3.6 × 2.2 cm. 1914.680 © The Cleveland Museum of Art, 1998. Gift of the John Huntington Art and Polytechnic Trust

Egyptian painting gives a very clear and complete account of life and afterlife. Frontal versus profile representation of figures was strictly codified for over two millennia. Artists rarely attempted spatial representation or foreshortening techniques, although we know that they were capable of achieving them, as they made brief appearances.

Naturalism was particularly developed when Pharaoh Akhenaten (*c.* 1370 B.C.) imposed monotheism, which lasted only one generation. Such an example is a remarkable illustration of painters' quick response to sudden change in millennial conventions, indicating that artistic style is more a matter of intention than of aptitude.

As in Egypt, early Greek figurative works consisted essentially of linear contours filled in with flat colour, whereas later paintings reflect a much higher degree of spatial organization.

Few mural or panel paintings from Ancient Greece survive, but we know that they existed. Our information is based on copies by the Romans, who increased the illusionistic quality of images – applying several layers of paint, one on top of the other. Their fresco technique became very elaborate, including mixtures of quicklime, chalk and soap applied on a dry surface, which was polished with a marble cylinder.

The Romans developed an intuitive form of perspective representation, with naturalistic effects enhanced by *trompe-l'œil* devices – by painting the canvas edges as 'window frames' through which the viewer looks upon a real scene. Pliny wrote a full account of the artistic techniques known until his time, which later guided the Renaissance painters.

Technique / Period	Size	Shade	Texture	Atmosphere	Linearity
Prehistory	+	+/-	-	-	-
Ancient Egypt	-	-	-	-	-
Ancient Greece	+/-	-	-	+/-	+/-
Ancient Rome	+	+	+	-	+/-
Renaissance	+	+	+	+/-	+

PERSPECTIVE AS PRACTISED IN VARIOUS PERIODS

The eye does not record images in three dimensions. This process is worked out by the brain. Prehistoric painters seem to have had notions of curvilinear perspective. According to an ancient Assyrian inscription, the apparent diminution of objects with increasing distance was described in the seventh century B.C.: 'Being carried up by an eagle to the heavenly throne of the goddess Ishtar, I was so clear-headed that I could observe that the earth became smaller and smaller until it reached a vanishing point.'

Painting without paint

For thousand of years, the content of a work of art was strictly codified, therefore the artist concentrated on the medium. New techniques were often instrumental in bringing about artistic developments.

At Fayum in Egypt, funerary customs in Roman times required the production of vivid likenesses. Amazingly fresh portraits were executed with a technique called encaustic. Dry pigment suspended in hot beeswax and resin gave a translucent and brilliant effect after polishing. (Wax was also used for burial masks.) The translucence and durability of this medium, and its slow-drying properties, made it very popular: encaustic portraits were produced in large numbers.

A unique form of portraiture, the staring icon ('image', in Greek), was inspired by the use of this technique and brought about a profound change in painting. When Rome declined, Byzantium took over as the centre for Christian art, dominated by the iconic style. Compositions were greatly simplified, modelling became harsher, and three-dimensional illusion all but disappeared.

Since abstract models were generally unavailable at the time, these compositions cannot be interpreted as regression. They were, rather, a form of symbolic expression emphasizing a renewed spirituality. This lasted over one thousand years, during which there was no science to speak of in the Western world.

PORTRAIT OF A BOY, PROBABLY FAYUM, EGYPT, SECOND CENTURY A.D.

Encaustic was used by the Ancient Greeks to decorate and protect their ships. Roman encaustic portraits had a stylistic influence on Byzantine painting and mosaics.

DETAIL OF *CHRIST BETWEEN SAINT PETER AND SAINT JAMES MAJOR*, CIMABUE SCHOOL, LATE THIRTEENTH CENTURY

In Byzantine style, canons of order and gesture, although not mathematical, were at work. The Italian master Cimabue was still influenced by the iconic style almost a thousand years after it first appeared.

The battle of Alexander against the Persians,
MOSAIC, FIRST CENTURY A.D.
This large stone and glass mosaic floor found in Pompeii is a copy of a Hellenistic painting (*c.* 300 B.C.). The sense of movement and crowding, as well as the excitement on Darius' (right) and Alexander's (left) faces, give an idea of the representational skills of ancient artists.
Museo Archeologico Nazionale, Naples

MANUFACTURE OF A STAINED-GLASS WINDOW
Highly architectured stained-glass windows are made of hundreds of small pieces of coloured glass. This image shows a craftsman preparing lead strips for assembling the pieces.

Christian painting was for a long period a forbidden activity secretly practised in catacombs or in small format – the latter possibly paving the way for a new genre, the miniature. Small transportable formats using metal and gems were likely precursors of mosaics and stained-glass windows.

Mosaics were built into the architecture they embellish, as an alternative form of painting. The Mesopotamians had used them to decorate their palaces, pressing pebbles into plaster to create an attractive floor. Natural stones, which were rarely bright, were used to cover floors. They were later replaced by cut pieces of marble tesserae (square tiles) for which the Romans substituted coloured reflective glass.

After the image was drawn on a plaster surface, the mosaics were pressed into position and angled slightly so that the light would glance off them, creating a scintillating effect. In Byzantine mosaics, the effect was augmented by a shimmering gold-leaf background, extending the walls and ceilings. The sophisticated use of mosaic in this art was one of unprecedented glory.

In the North, stained-glass windows, which were to become the leading visual art form, were mentioned as early as the sixth century. This art reached maturity towards the end of the Middle Ages, by which time regional schools led by monasteries were competing internationally; colour recipes proliferated as a result of widespread chemical experimentation.

Orchestrating the construction of a large stained-glass window was no simple matter. The designers were constrained by the size limits imposed by contemporary glass-production methods. In coloured glass, made to order, segments were first placed over the artist's drawing and fine detail painted on them in black enamel. The window was then assembled with lead strips that were fused to the glass by firing. The dark and heavy lead contours accentuated the schematic rigour of the design and made colours look even brighter.

Cathedrals, those 'Encyclopedias in Stone and Glass', were not only places of worship but also replacements for books in the education of the illiterate. Mosaics and stained-glass windows were gradually forgotten and re-emerged in the wake of Art Nouveau, in the nineteenth century.

Painting without a frame

The Jingting Mountains in the Fall, Shitao, seventeenth century

In Chinese painting, the human silhouette is lost in the vast landscape. Size is often related to status rather than to the laws of perspective. The painter deciphered nature as an ideogram: behind the image's poetry lies a philosophical choice.

Musée National des Arts Asiatiques-Guimet

The contrast between Eastern and Western ways of thinking about man and the universe is readily summed up in painting. For the Chinese, landscape was an elevated genre. The viewer was projected into it as an integral part of nature.

Brush and ink painting date back to 1000 B.C. Standards in China were set as early as the sixth century A.D., when six laws, which governed painting until recently, were established. They concerned how to hold and use a brush, how to select appropriate colours, rules of composition and training, and subjects considered worthy of the painter's art.

Academies were founded, whose principles were widely respected throughout the Far East. The canons were carefully preserved, yet – as in Byzantine style – within the canonical form, there was abundant room for subtle experimentation with detail, colour and feelings.

Painting was a spiritual exercise in which the artist's main goal was to create a focus for meditation. The time the artist devoted to studying nature's details was carefully balanced with the effort dedicated to actually executing the work.

The Chinese, who were the first to see in painting a higher art form rather than a craft, pursued it by scrutinizing nature, almost like scientists. The philosopher and painter Guo Xi (eleventh century) wrote with a touch of anthropomorphism: 'Water courses are the arteries of a mountain; grass and trees its hair; mist and haze its complexion.'

Far Eastern painters knew about linear perspective but since the viewer was considered to be 'inside' the painting, the illusory effect was irrelevant. They tended to employ aerial perspective, making distant objects look increasingly blurred. Their painting suggested a sense of depth even though linear perspective was not used.

Throughout the East, original styles developed. For example, Persian painting depicted exquisite decors on flat surfaces. In the sixteenth century, the Japanese opted for uniform flat colour planes, a style called ukiyo-e. They relied mainly on the size of overlapping 'floating' forms to suggest space, which would greatly inspire Degas, Toulouse-Lautrec and Gauguin.

Nearly one century before Cézanne painted his multiple views of *La Montagne Sainte-Victoire,* Hokusai (1760–1849)

painted thirty-six different views of Mount Fuji. In his own way, one of the most popular of Japanese painters questioned the principle of a unique viewpoint.

As we reach the third millennium – the West, so-called guardian of science, meets the East, holder of ancestral values – aesthetic and philosophical convergence is originating new creative forms.

DETAIL OF A JAPANESE PAINTING, NINETEENTH CENTURY
This educational picture represents different foetal positions during pregnancy. The Japanese theatre of life portrays intimate scenes, in which nude bodies appear as natural subjects of contemplation.
National Library of Medicine, Bethesda

STAR MANDALA, EDO PERIOD, JAPAN, 1615–1867
Asian artists represented the mandala, a Sanskrit word meaning 'magic circle' – whose shapes are spherical or radially oriented. Such diagrams of the cosmos included concepts of geometry, geography and numerology. Analogous representations exist in Christian art through Christ and the four apostles; the cross, indicating the cardinal points, had a cosmological function. The Swiss psychiatrist Carl Gustav Jung found the mandala symbolism in his patients' dreams.
Philadelphia Museum of Art. A Gift of the Friends of the Philadelphia Museum of Art

The illusion of reality

While stained-glass windows reached their apogee in France towards the end of the thirteenth century, in Southern Europe mosaic was replaced by painting partly because of cost. In the wake of Gothic cathedral building, mural painting found fertile ground in Italy where windows were kept small because of the hot sun, leaving large wall surfaces to fill.

Confronted with the Gothic style, dignified Byzantine representation came to an end. Giotto (1266–1337) broke with tradition by introducing the illusion of spacial recession in painting for the first time in centuries; he reinvented the means of creating depth on a flat surface. A shepherd in his youth, and a keen observer of his surroundings, Giotto found his way empirically. He was able to represent three-dimensional space and volumes occurring in nature, despite the fact that perspective and anatomy would not be developed for at least another century.

Skilful in many arts, Giotto was mainly active as a fresco designer. Despite the restrictions of this technique, he managed to introduce radical new expressions of space and movement. As a symbolic turning point for painters, he was named headmaster of the Florence Cathedral workshop, where he designed the bell tower, a privilege until then reserved for sculptors and architects. The multi-talented Giotto experimented with various concepts and media.

The Slaughter of the Innocents,
Fresco by Giotto, 1303–5
Each scene of Giotto's series seems to be painted for a single observer, placed at a specific spot. Harmonious volumes and proportions, single figures versus tightly clustered groups, and a unique balance between the depicted architectural setting and narrative drama, were among the characteristics of his painting style.
Capella Degli Scrovegni, Padua. Commune di Padova Settore Musei e Biblioteche

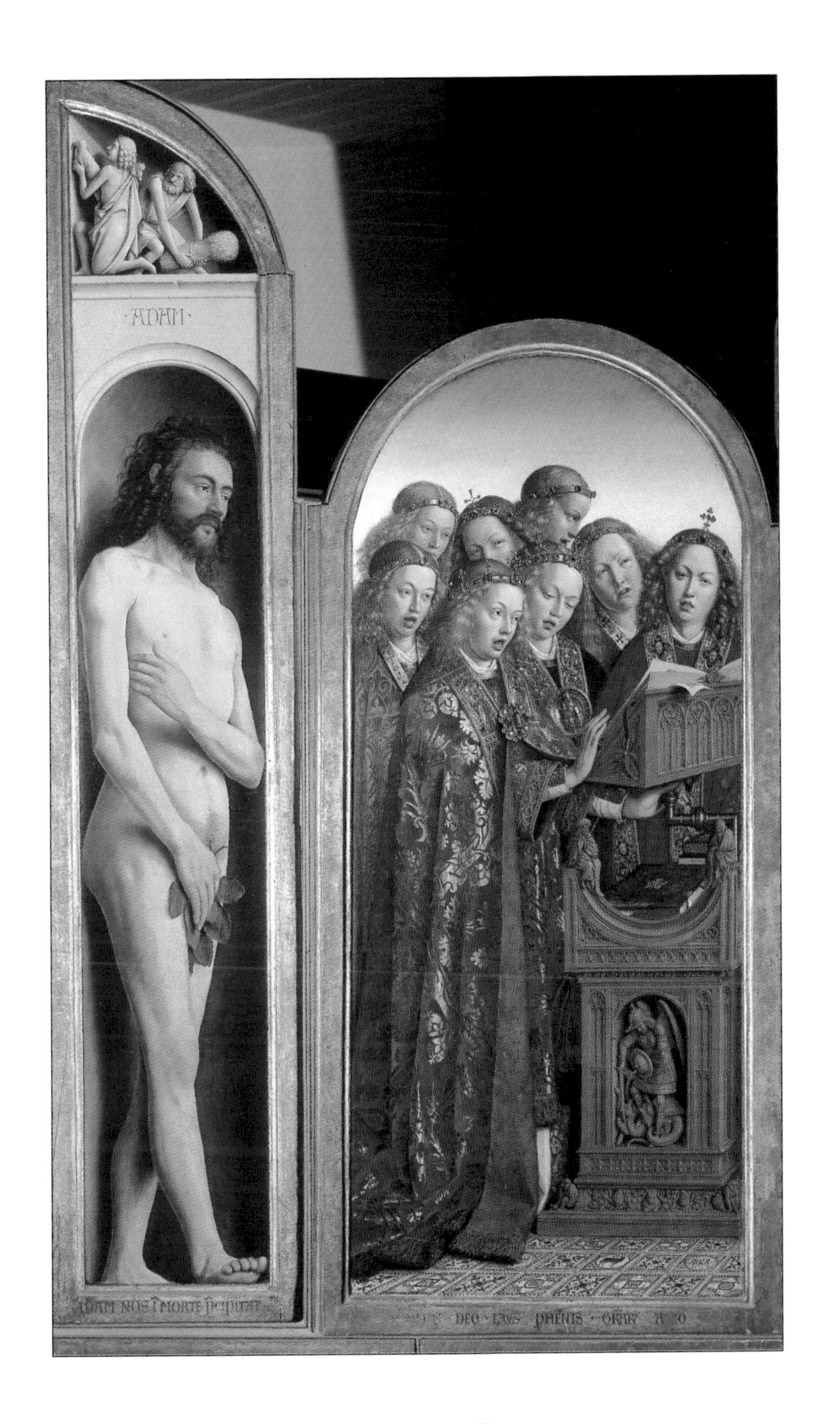

DETAIL OF THE ALTARPIECE, JAN AND HUBERT VAN EYCK,
SAINT BAVO CATHEDRAL, GHENT, FIFTEENTH CENTURY

Northern artists experimenting with oil paint reached an unprecedented degree of naturalism. The Van Eyck brothers pioneered some of the earliest nearly life-size nudes. The self-portrait, too, came of age in oils.

Photo © Paul M. R. Maeyaert

Although the story of painting cannot simply be described as a succession of techniques, the earliest Renaissance art treatise was just a compilation of recipes. Painting was perceived as a cumulative discipline aimed at realistic representation.

Artists experimented with different materials through the ages – plaster, encaustic, glass or marble tesserae – and systematically succeeded in using these techniques in innovatory ways.

Of all the mediums that image-makers experimented with, oil paint is the one that most fundamentally transformed their art. It made it possible for the painter to change his mind at will, and has remained the basic medium ever since. Oil paint allowed the artist to work slowly and permitted smooth blending of colours. Since then, painting has acquired a new dimension. The small, transportable easel painting which is proper to Western civilizations has almost become a symbol of independence.

The Flemish were the first to use oil paint consistently. In the fifteenth century, painters were often chemists preparing their own pigments, grinding them in a mortar and mixing them with a binding agent to make a paste. One of these was oil essayed in various proportions. Preferred for its slow-drying quality, it was finally adopted itself as the pigment suspension agent.

Oil paint was applied more effectively over a dark background. It became possible to work layers of pigments like glazes, so that brushstrokes were invisible, and to obtain a variable degree of opacity or translucency. The finest details of form, texture and light could be rendered with unprecedented fidelity, which particularly appealed to the Flemish sense of narration. In Flanders, painting, like polyphonic music, was reaching new heights.

Constructing a painting from a myriad of details that coalesce into a space was a new method of composition that inspired the following words of the art historian Erwin Panofsky (1892–1968): 'Jan Van Eyck's eye operates as a microscope and as a telescope at the same time – and it is amusing to think that both these instruments were to be invented some 175 years later, in the Netherlands – so that the beholder is compelled to oscillate between a position reasonably far from the picture and many positions very close to it.'

From craft to art

In Italy, painters still worked according to the antique method of covering walls and panels with layers of fresh plaster. The painter drew the design with charcoal; scratching was carried out following the contours, which defined the colours to be applied. They were worked on the white background and were not very intense, because of the bleaching effect. Since plaster can not be re-wetted once it is dry, the artist never applied more than he knew he could cover in a single day, leaving marks on frescoes where plaster joints occur.

Departing from ancient techniques, the Renaissance artists were set to create a totally new approach to painting. Viewed thus far as a time-honoured craft, it became a true art in Italy, as soon as scientific concepts were applied to it. The architect Leon Battista Alberti wrote a treatise on painting, *De Pictura*, (1436) which became the cornerstone of Western painting for the next five hundred years.

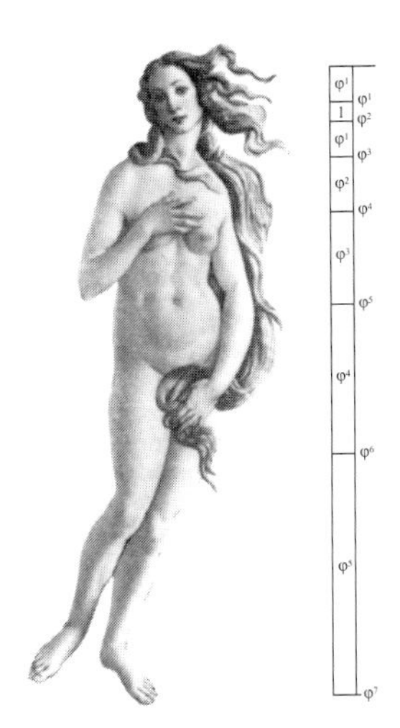

BOTTICELLI'S *VENUS* ANALYSED ACCORDING TO THE GOLDEN SECTION BY THEODORE COOK, 1914
Botticelli's *Venus* was one the first monumental images of a nude goddess since Roman times. Trained in optics and geometry, he developed an idealized style based on mathematical proportions. Later paintings by Velázquez or Delacroix also contain references to the Golden Section, and the Cubist journal was called *La section d'or.*
The Curves of Life. *Dover Publications, Inc.*

Renaissance painters aimed at perfecting the illusion of a real world on a flat surface through modelling form with pigment, on the one hand, and perspective construction, on the other. The philosophical dimension was at least as important as the technical one. Many symbols in paintings indicate that artists were indeed familiar with philosophy.

Perspective responded to a need for exactitude and predictability. This can be detected in the works by Piero della Francesca, the first painter who used perspective as a science completely determining the painting. When Piero, who wrote on optics and mathematics, drew a body or a monument, he designed them seemingly unemotionally as a combination of basic geometrical forms: pyramids, cones, cubes and spheres.

Whereas Kepler's astronomy was still largely based on intuition and aesthetic pursuit, Piero modulated light and shadow with great insight, using 'quasi-Newtonian' methods. In science, a systematic approach to problem solving was to appear two hundred years after Piero's.

ARCHITECTURAL PROJECT FOR AN IDEAL CITY, ATTRIBUTED TO SANGALLO GIULIANO GIAMBERTI (FORMERLY TO PIERO DELLA FRANCESCA), LATE FIFTEENTH CENTURY
Italian Renaissance art was an attempt to reconcile aesthetic and scientific concepts. Piero measured over one hundred points on a single object before painting it in perspective.
Galleria Nazionale delle Marche, Urbino. Alinari-Giraudon

Then, two styles dominated successively. The high Renaissance, as represented by Raphael's (1483–1520) idealizing Classicism, and Mannerism, as illustrated by Michelangelo's (1475–1564) *Last Judgement,* a masterful combination of anxiety and virtuosity. Leonardo, their contemporary, who basically was a researcher, wrote: 'Art truly is a Science.'

Driven by metaphysical and physical curiosity, Leonardo experimented with composition, volume and movement, and techniques such as blurring of outlines, or *sfumato* – literally, turned to vapour. He was testing all sorts of materials, with sometimes disastrous effects.

Despite this research, painting was classed with the mechanical arts and the craft-like connection was still present. Leonardo the architect-engineer, Michelangelo, the sculptor-architect and talented poet, was also a practical engineer when needed (he designed the Florence fortification).

Leonardo and Michelangelo did not separate their fine art work from their other inventions. Both geniuses must have confronted tormenting decisions about what to focus on. Raphael, the painter-architect, fully integrated the techniques relating to anatomy and perspective. Clarity and harmony dominated.

Most of their masterpieces were achieved within just a few decades: suddenly art had taken over. Like modern scientists, painters who no longer were craftsmen looked ahead, driven by a profound need to herald change.

THE SCHOOL OF ATHENS, RAPHAEL, 1510–11

This painting, considered as one of Raphael's masterpieces, represents a group of philosophers gathered around Plato and Aristotle. Each one is characterized individually: Plato points upwards, suggesting an abstract philosophical approach to knowledge, whereas Aristotle gestures towards the earth, indicating his preference for concrete facts derived from the observation of nature. Euclid is seen bending over a compass and board (right foreground). The figure posed as a prophet may well be a homage to the ageing Michelangelo, whereas Plato bears a resemblance to Leonardo. Raphael has portrayed himself staring at the viewer, on the right side of the composition.

Courtesy of the Vatican Museums. Photo P. Zigrossi. MAG.1996

Depth cues obtained by shading

If the tops of the circles are light, they look like bumps. If their bottoms are light, they look like depressions. Painters master such visual effects, intuitively. Cognitive scientists analyse them in order to understand how depth is reconstructed by the brain.

Not surprisingly, artists in the cosmopolitan city of Venice lost little time in adopting the oil medium, introduced in the sixteenth century by visiting artists from the North.

Brilliant primary colours and asymmetrical composition were pioneered to produce lively movements. Using bold brushstrokes, Titian (1488–1576) outlined luminous and voluminous forms, modelling them sculpturally with thick daubs, a technique called impasto (meaning thick laying on of colour). This change in technique meant, as usual, a change in intent.

Painting became an integral part of architecture, and baroque architecture resembled painting. Artists seized the emerging technical opportunities to communicate their religious spirit. The work of Michelangelo Caravaggio (1573–1610), constructed from dramatic passages of light, *chiaro*, and shade, *scuro* – including *virtuoso* incidents of foreshortening that projected forms out into the viewer's space – led to a revolution in religious art.

The light source in Caravaggio's paintings seemed to come from outside the depicted scene: from the real world. His models were ordinary people whose features he copied to portray saints. While the Church despised him, his art spread throughout Europe as his imitators, the *Caravaggisti,* dispersed to decorate churches and buildings in France, Spain and the Netherlands.

David with the Head of Goliath,
Caravaggio, *c.* 1605–6

Caravaggio heightened the dramatic content of painting by intensifying contrast. Baroque painters were masters in psychological build-up.

Galleria Borghese, Rome. Archivio Fotografico Soprintendenza Beni Artistici e Storici di Roma

In the Netherlands, the wealthy bourgeoisie bought still lifes, landscapes and *genre* paintings that closely represented the world around them, or which depicted exotic New World flora and fauna. These works were executed with a great sense of observation corresponding to the growing confidence in science.

Rembrandt van Rijn (1606–69), the best-known Dutch painter, stuck to more traditional themes, however. The Bible and portraiture provided him with ample subject matter for his art – a labour of passionate experimentation rather than accurate description.

He broadened his creative potential by technical explorations, which can be traced through dozens of self-portraits. He obtained unique effects, using unusual calcareous mixtures for white, and glass powders for blue. His rare oils were blended in complex recipes which not even his pupils could reproduce.

Rembrandt's masterpieces were subject to intense preliminary research with almost any kind of medium. To traditional tools such as charcoal, lead pencil, chalk, ink, wash or watercolour, he added forgotten techniques, and in particular, he took etching to new heights. He is most widely known for his deeply moving oil paintings, which were finally composed of strokes so thick they could have been applied directly with his fingers. X-rays of his paintings indicate that he often reworked them. Like an inventive scientist, Rembrandt restlessly researched means of expressing deeper emotion.

A change in taste gradually occurred, opposing an earlier colourful trend to one of monochromy dominated by black and white contrast. From here on, it became difficult to imagine the antique world in bright colours. The desire to represent the material world as realistically as possible prompted painters to use instruments of measurement. They also used small deceptions: for example, careful examination of paintings by Jan Vermeer (1632–75) reveal the existence of a tiny hole where perspective lines converge. (This was where the artist fixed a pin attached to a thread dipped in white powder. By stretching and flicking it, the painter obtained the guidelines for his depicted spacial recessions.)

In composing his painted interiors bathed in meticulously depicted light, Vermeer probably used the camera obscura. In 1622, the Dutch diplomat, poet and art connoisseur, Constantin Huygens, warned that this instrument would cause the imminent death of painting! Just like photography was supposed to do, a century later.

The Anatomy Lesson of Dr Tulp, Rembrandt, 1632

Artists have always shown interest in scientific subjects – seventeenth-century painters in particular. Rembrandt painted this masterpiece when he was 26.

169.5 × 216.5 cm. © Mauritshuis, The Hague, the Netherlands

Samuel Van Hoogstraten's *Peepshow*, seventeenth century

Working in Rembrandt's studio, Van Hoogstraten used the above perspective box.

Reproduced by courtesy of the Trustees of The National Gallery, London. The National Gallery Picture Library

Portrait of Antoni Van Leeuwenhook, Jan Verkolje, seventeenth century

In the course of his work as a draper, Leeuwenhook closely examined textiles with a magnifying lense; he invented the microscope. Leeuwenhook was appointed trustee of Vermeer's bankrupt estate after the latter died. According to legend, Leeuwenhook inspired Vermeer's *Geographer* and *Astronomer.*

Collectie Rijksmuseum. © Rijksmuseum-Stichting, Amsterdam

Thomas Gainsborough's Show Box, 1781–2

Artists endulged in technical gimmicks as tools for creative purposes. Dutch painters used the camera obscura as did Gainsborough, in England, and Canaletto, in Italy. Others used cylindrical mirrors to deform images.

© The Board of Trustees of the Victoria & Albert Museum. V&A Picture Library, London

Exact sciences were increasingly believed to comprehend God's design and the divine laws of nature. Landscape painting gradually became recognized as an independent genre rather than as background to mythological scenes.

The French painter Claude Lorrain (1600–82) used mirrors to help him compose luminous landscapes; so did Nicolas Poussin (1593–1695), who symbolized French rationalism even more, although he too spent most of his career in Italy. Poussin intensely studied optics and mathematics while analysing the effect of light and said: 'colours in painting are as allurements for persuading the eyes as the sweetness of meter is in geometry.'

Heated debates on the importance of line versus colour were raging. The Académie des Beaux Arts had been founded to support painters' mechanical training with theoretical knowledge, as painting was upgraded and classed with the liberal arts. But, progressively, the Academy's influence became stifling.

A century later, Jacques Louis David (1748–1825), aiming for extreme neo-classicism, imposed drawing, naturally, as the dominant element of painting. In this context, the strongest supporters of the artist's freedom of expression were scientists such as Denis Diderot (1713–84), the first art critic. By then, the artist had become a definite generator of ideas rather than a manipulator of rules and recipes.

CLAUDE GLASS, UNSIGNED, LATE EIGHTEENTH CENTURY
Poussin staged small wax figures on which he based his compositions. Although we do not know for sure that Lorrain used the above dark curved mirror to reduce the tonality of light, most of his imitators did. The Claude mirror became as popular in the eighteenth century as photography in the nineteenth.
By permission of the Trustees of the National Museums of Scotland

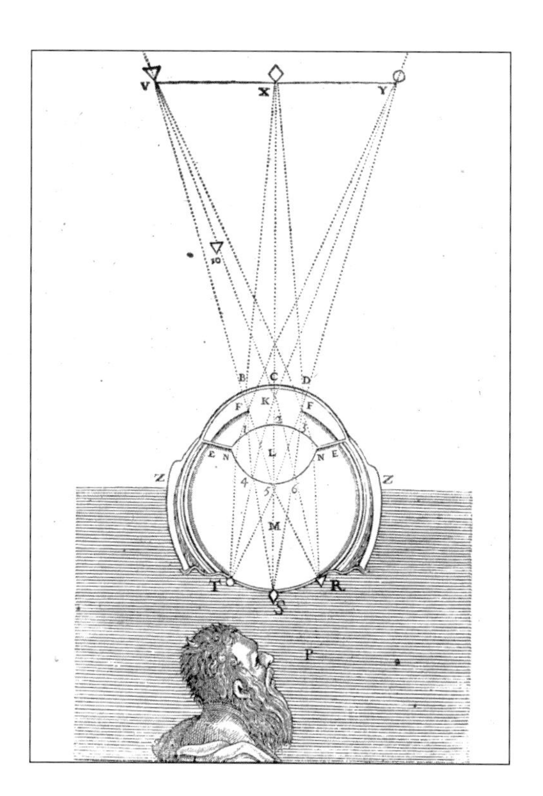

LA DIOPTRIQUE, RENÉ DESCARTES, LEIDEN, 1637
The idea that light, the primary vehicle of observation, is made of particles projected into the eye is traced back to Pythagoras; instead, Plato taught that light was emitted from the eye. Descartes showed that images are inverted on the retina; from there onwards, the interest in optics extended to eye physiology.
Department of Special Collections. University Research Library, UCLA

Oath of the Horatii, Jacques Louis David, 1784
David's figures, as solid as statues frozen on the picture plane, became a symbol of ideological correctness when the French Revolution broke out. The reaction to his work at the time was comparable to that evoked by Picasso's *Guernica* in the twentieth century.

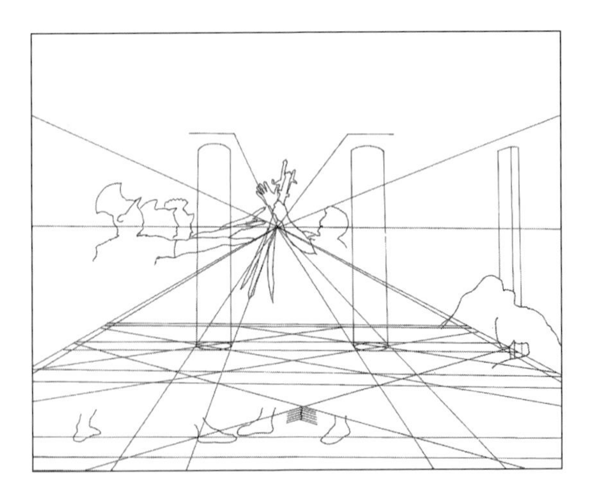

Perspective analysed by Martin Kemp
The Science of Art. *Yale University Press. Pelican History of Art*

About relativity

Representational painting reached both a summit and an impasse of mannered perfection. Where to go from there? Innovation came from England, where John Constable (1776–1837) spent his life studying meteorology – then a fashionable science – on local landscape in order to render its forms in all their luminous subtlety.

More successful in France than in his own country, proceeding stepwise by trial and error, he stated: 'Painting is a science, and should be pursued as an inquiry into the laws of nature. Why, then, may not landscape be considered as a branch of natural philosophy, of which pictures are but experiments?'

His bold contemporary, William Turner (1775–1851), went further still, in his late atmospheric studies of roaring seas, stormy skies and smoke-filled air that became harbingers of Impressionism and even of abstract art.

Theoretical arguments on the function of the retina and the physical nature of light itself were fashionable. Goethe's theories were pitted against those of Newton, which were at last gaining acceptance. Helmholtz's *Manual on Physiological Optics* (1856), a theory of colour and tone sensation, contributed meaningfully to the fray, which saw science and art criticism rally around the same issues and even share a common terminology (such as chromatic aberration, prismatic decomposition).

The painters themselves had already made the next step: the expression *'l'art pour l'art'* was coined in 1835. Édouard Manet (1832–83) cast off several hundreds of years of aesthetic convention by abandoning naturalism in favour of a stylized

STUDIES ON REFLECTION AND REFRACTION IN GLASS SPHERES CONTAINING WATER, WILLIAM TURNER, 1815
Turner opened new paths for the Impressionists to explore further. Although painters had hundreds of synthetic hues to work with, many of them proved fugitive in the extreme. Some nineteenth-century yellows and pinks are so degraded that we can only guess at their original appearance.

D 40025 CXCV 177C. Tate Gallery. Tate Picture Library, London, 1998

realism loaded with disquieting socio-political content. The flattening of form and space that he and his peers effected was also inspired by the Japanese art display at the Paris World Exhibition.

Olympia, Édouard Manet, 1863
The French painter, Manet, introduced subjects considered so far unworthy of high art. Here, a courtesan looks us straight in the eye.
Musée d'Orsay. © Photo RMN. Gérard Blot

While Manet compressed space and dispensed with allegory, Monet (1840–1926) set out to record the transitory retinal verities of the here and now by closely observing and depicting the same scene – *Haystacks, Rouen Cathedral* – at different times of day. His eye functioned like a camera: he painted 'what he saw rather than what he knew'. Form fragmented into its colour constituents as light dissolved what it defined.

In 1841, the collapsible tube, which became available commercially, liberated the painter from the necessity of returning to the studio to paint from sketches made outdoors. Artists would chase hues as their predecessors tracked details. Tube

Claude Monet dans son Bateau-Atelier, Édouard Manet, 1876
While Manet flattened space, Monet emphasized the fleeting nature of time, recording the optical effect of changing light. A keen nature observer, and expert in botany, he had this boat designed especially as a floating 'laboratory'. Impressionists reverted to the use of white underpaint, to better reflect the shimmering qualities of light.
Neue Pinakothek, Munich.
Bayerischen Staatsgemäldesammlungen

paint was thick, and colours could be mixed directly on the canvas. They tucked their easels under their arms and went out into the countryside to study nature.

The rising artistic focus on colour relationships coincided with major discoveries by scientists, Gustav Kirchhoff (1824–87) and Robert Bunsen (1811–99). A new tool was used to analyse the colour spectrum of light produced by burning elements (the spectrophotometer, 1859). The results led to the discovery of other elements; the first one found by spectroscopy was caesium, which means 'sky blue', referring to the colour of its spectrum, the next one was rubidium . . . then thallium was named after the colour of grass.

By 1863, the year of the *Salon des refusés,* scientists catalogued the spectra emitted by the elements, which revolutionized ideas about their internal structure. It was thus colour analysis that pointed the direction of modern science.

The wide range of synthetic colours particularly appealed to the Post-Impressionists, who constructed images from tiny dots and daubs; their canvas surfaces seemed to mimic the function of giant retina cells, acting as light receptors and colour decoders.

The technique was anything but haphazard. The French painter Georges Seurat (1859–91), dubbed 'the little chemist' by disdainful critics, said: 'Taking for granted the phenomena

MODELS, GEORGES SEURAT, 1886–88

On the wall of his studio depicted in this painting, hangs Seurat's major achievement, *La Grande Jatte,* for which he made dozens of preliminary studies. Seurat died at age 32 and, not surprisingly, left few paintings.

Oil on canvas, 78 3/4" × 98 3/8" BF inv. number 811, Main Gallery.

of the duration of light impression upon the retina: a synthesis follows as a result. The means of expression is the optical mixture of tones and colours and their reactions according to the laws of contrast, sequence and irradiation.'

Following the climax of 'scientific painting', the intense expressionism of Vincent van Gogh (1853–90) and the private symbolism of Paul Gauguin (1848–1903) had their counterparts in emerging sciences, such as sociology and psychiatry.

Although still in its infancy, neurophysiology was attracting much interest among scientists, one of whom, Sigmund Freud, began to study the brain and its relation to physical functions. Specialists assumed that mental illnesses seated in different parts of the brain were expressed in the physionomy.

Some of Freud's most insightful studies were concerned with the neuroses of highly creative artists. He particularly liked to refer to: 'this light narcosis, art' and considered that: 'this staging of illusions, that we recognize as such . . . is probably the apex of the joys of imagination.'

Freud could conceivably have been a significant contributor to different areas of knowledge. As a young man, he

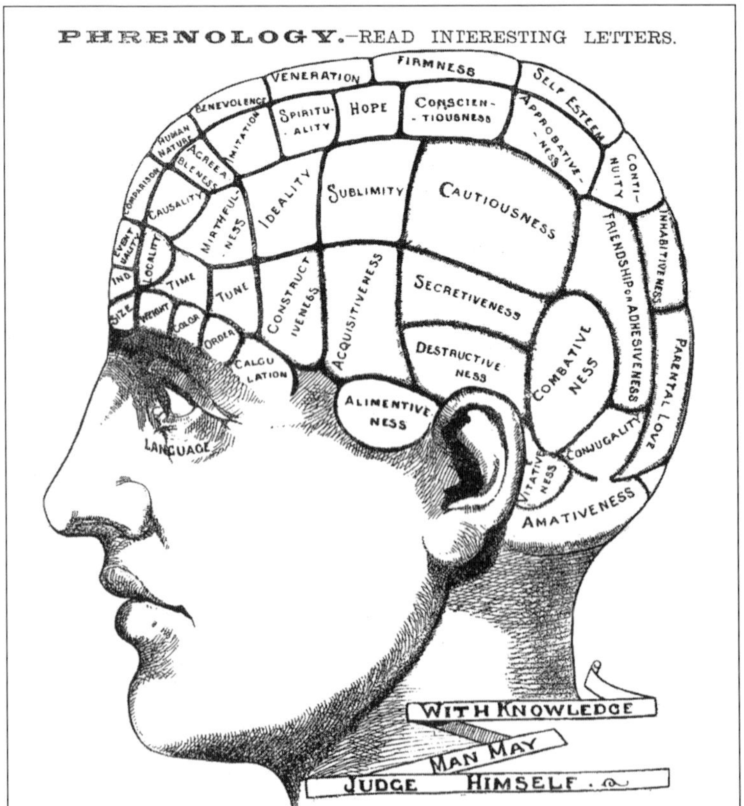

ANATOMICAL MAP OF THE BRAIN, NINETEENTH CENTURY

Paul Broca convincingly demonstrated a link between speech and a specific area of the brain – by discovering a lesion during an autopsy on a man who could not speak intelligibly. The art historian Panofsky considered that 'the study of the proportions of faces and bodies was very useful in the development of new sciences such as anthropology'. Degas incorporated such concepts in his art work.

© Hulton Getty/Fotogram-Stone Images, Paris

suppressed his strong cultural streaks, embracing the medical field inspired by . . . Goethe's research. Freud's major contribution, psychoanalysis, had been somewhat anticipated a century earlier by the artist William Blake.

The mystery and fascination of the subconscious gave impetus to the Surrealist movement which was initiated by writers. Their leader, André Breton (1896–1966), a medical student shaken by the First World War, was familiar with Freud's work. Surrealism was described as 'thought dictated in the absence of control exerted by reason, and outside all aesthetic or moral preoccupations'. Free association, a technique used in psychoanalysis, was transferred to the process of art-making and labelled 'automatism'.

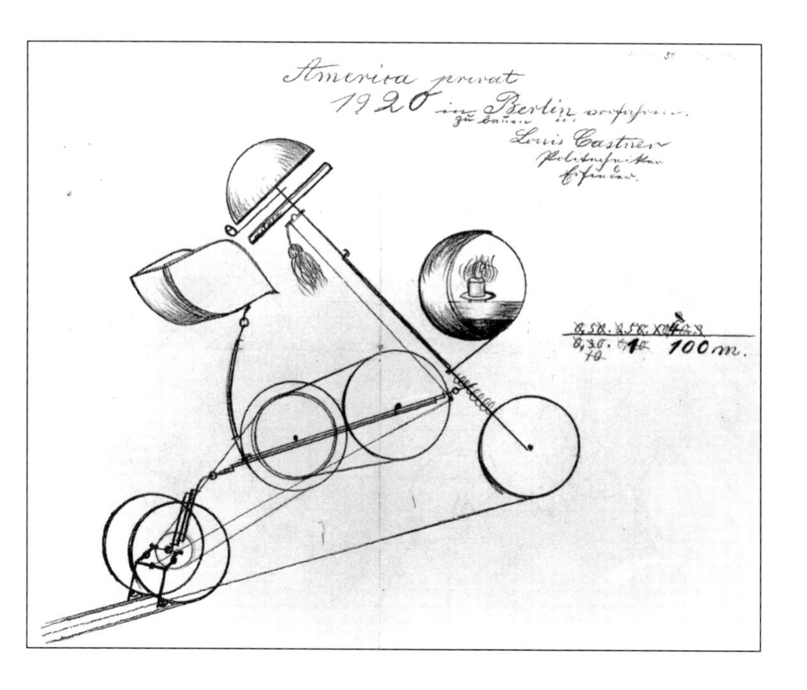

AMERICA PRIVAT, LOUIS CASTNER, *C.* 1920
Jean Martin Charcot, a French physician who gathered a photographic collection of his mentally ill patients, greatly inspired Freud. The art work by psychiatric patients collected by Hans Prinzhorn is equally remarkable for its high quality. It inspired German Expressionists, Abstract and Surrealist painters.
Indelible pencil on paper, case number 251 Inv. 3116 verso.
© Prinzhorn-Sammlung der Psychiatrischen Universitätsklinik Heidelberg. Photo Klinger

Similarly, Henri Poincaré, who presaged the theory of relativity, described the steps of his own creative process: 'there is the initial period of conscious work dealing with the problem, then, a period during which the unconscious mind seems to be active when an appropriate hypothesis strikes the thinker with its aesthetic properties much as a good work of art does. Proof has next to be worked out.'

At the time, numerous artists and scientists acknowledged their dreams in terms of visual symbols that suddenly revealed the unconscious mind and boosted their creativity.

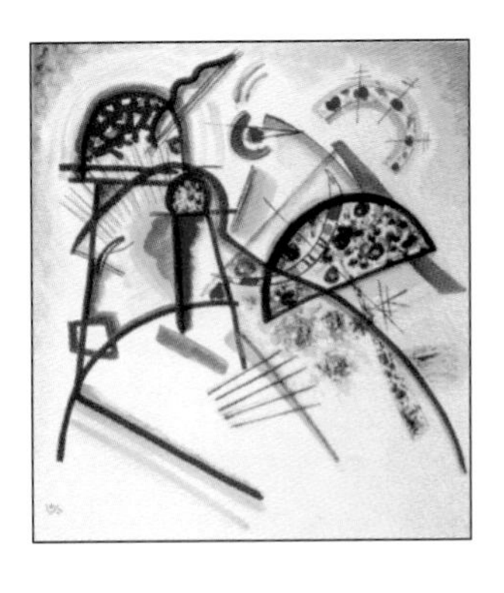

UNTITLED WATERCOLOUR,
WASSILY KANDINSKY, 1923

Graphisches Sammlung Albertina, Vienna

SUPREMATIST COMPOSITION: WHITE ON WHITE,
KASIMIR MALEVICH, 1918

One of Malevich's paintings, *Black quadrilateral*, symbolizes the eclipse of the sun in Western painting, establishing the supremacy of a new pictorial order. His radical simplicity matched Einstein's theory of relativity. By redefining the environment, both totally surprised their respective audiences.

Oil on canvas, 31 1/4" x 31 1/4". The Museum of Modern Art.

DARK OVER BROWN, NO. 14, MARK ROTHKO, 1963

The Abstract Expressionist painter, Mark Rothko, used similar patterns to convey his spiritual angst. Perhaps future generations will interpret such paintings as allusions to a post-nuclear void.

Collection Musée National d'Art Moderne/Cci/Centre Georges-Pompidou. Photothèque des collections du Musée National d'Art Moderne/Cci/Centre Georges-Pompidou, Paris

Around the time that the Surrealists were bringing forth their dreams in images and proclaiming the superiority of the irrational vision, Wassily Kandinsky (1866–1944) was to reach a stage where painting and music – a non-representational and thus superior art form in his opinion – become a harmonious philosophical creation uttered in mathematical form.

Kandinsky, the father of abstract art, described some of the factors that underlay the development of his new style as 'an arrangement of abstract lines, shapes and colours'. . . . 'A scientific event cleared my way of one of the greatest impediments. This was the further division of the atom. The crumbling of the atom was to my soul like the crumbling of the whole world.'

Pure colour and the flat surface of the picture plane were explored by innovators such as Kasimir Malevich (1878–1935). Piet Mondrian (1872–1944) divided his own painting surfaces into grids of unmodulated colour inspired by mystical geometry.

He responded to a changing world order increasingly dominated by science: 'It is thus clear that man has not become a mechanic, but that the progress of science, of technique, of machinery, of life as a whole, has only made him into a living machine, capable of realizing in a pure manner the essence of art.'

Pioneers of abstract art were soon joined by an army of followers who continued to probe its optical and emotional limits. Total abstraction, like an end to all things, is an unprecedented chapter in the story of art.

Abstract art became a particularly interesting tool for analysing visual effectiveness: manipulating a limited number of variables in a painting foreshadowed a method used by cognitive scientists half a century later.

However, long before abstract art developed, a systematic study of space had already been pursued. Paul Cézanne (1839–1906) considered the aim of the Impressionists to capture the passing moment, an abandonment of the study of solid shapes found in nature. To one of his correspondents he gave the following advice: 'Treat nature by the cylinder, the sphere, the cone, everything in proper perspective so that each side of an object or a plane is directed towards a central point.'

Within this framework, Cézanne completely revised the visual means for indicating mass, space and gravity, establishing with colour an equilibrium between the representation of solid, three-dimensional objects and the flat painting surface. Scientists of the day were aware that classical geometry no longer met the challenge of the multi-dimensional space they were to tackle.

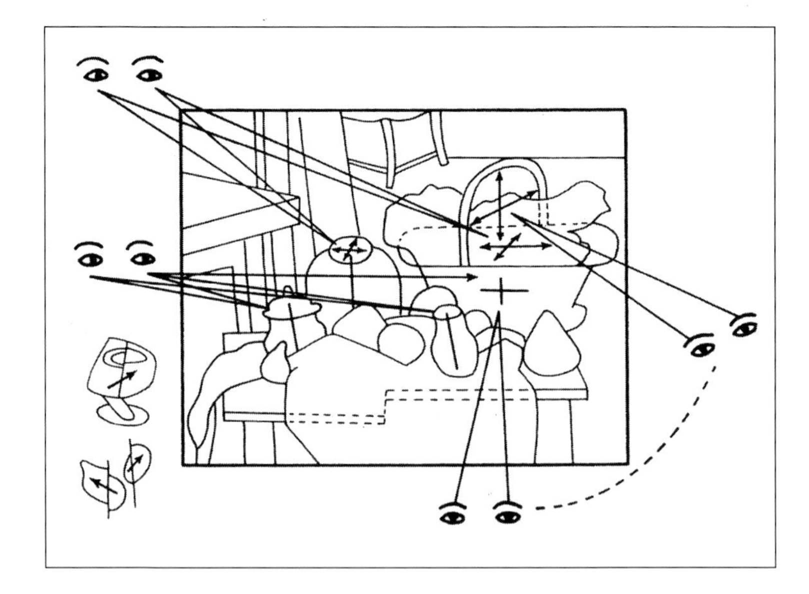

CÉZANNE'S COMPOSITION *STILL LIFE WITH FRUIT BASKET*, ANALYSED BY ERLE LORAN, 1943
This diagram shows how far from one-point perspective Cézanne chose to stray in order to incorporate several points of view into a single composition. From the visual evidence relating to space and gravity, we might assume that he was versed in modern physics and familiar with the theory of relativity, which of course he could not have been, since he died before Einstein's work was known to the public.

Cézanne's approach seemed like a pursuit of images that existed in his imagination. Of the mountain which he painted again and again, he said: 'I need to understand the geology, how Sainte-Victoire takes root, the colour of its earth. All this makes me a better man filled with emotion. It is bound to make me clear-sighted. . . . All things are linked. Understand me well, if my canvas is saturated with vague cosmic religiosity that makes me better, it will touch upon others' sensitivity. I must understand the geometry that keeps my reason straight.'

Perhaps one of the most significant messengers of twentieth-century science was Pablo Picasso (1881–1973), although he would have bristled at the idea. Similarly, Einstein was totally out of touch with modern painting and would have been surprised by perceived links between his work and Picasso's. Yet, today, it is hard to dismiss what seem like conceptual relationships emerging in different disciplines: art and science travelled on parallel lines.

Picasso painted 'what he thought rather than what he saw'. He viewed his painting as research and saw a logical sequence in his sketches, which he numbered and dated, for he firmly believed that some day they would be useful to those who study 'the science of creativity'. He considered that it is not sufficient to know artists' works, but that it is also necessary to know when they were made, why, and under what circumstances.

'I do not search, I find'. . . . What? Something that was already inscribed in his brain? The film by Henri-Georges Clouzot (1907–77), *Le Mystère Picasso,* powerfully illustrates the creative genius' fever. This document is even more impressive if we think of how little we understand about Picasso who lived among us for so long.

When looking at *Les Demoiselles d'Avignon,* it is easy to comprehend that Picasso said: 'In the old days, pictures went forward towards completion by stages. Every day brought something new. In my case, a picture is a sum of destructions.'

Despite the efforts of some creators to integrate art with science and technology, the world's values were smashed and some painters' work even seemed to allude to a potential nuclear disaster. Between two World Wars, the definite end-point to the beauty ideal and traditional values in art could be interpreted as an early warning sign of environmentalists.

NUDE DESCENDING A STAIRCASE, NO. 2,
MARCEL DUCHAMP, 1912

In this early work, Duchamp (1887-1968), anticipated Futurism. Also the champion of Dada, he soon abandoned painting for a mode of philosophical inquiry which resulted in the redefinition of the art object itself. Along the way, he tinkered with mechanical concepts and actual machines. He abandoned art to concentrate on chess, another intellectual challenge of which he was a master.

Oil on canvas 57 7/8" x 35 1/8". Philadelphia Museum of Art. Louise and Walter Arensberg Collection

Jackson Pollock, photo by Hans Namuth, Springs, Long Island, 1950
Fluid was projected on canvas by dripping or pedalling a bicycle. Such painting experiments resembled lessons in hydraulics and evoked violent sensations. Pollock became one of the most forceful and inventive figures among New York's Abstract Expressionists, whose art gave free reign to the subconscious.

In response, the painter expanded beyond traditional borders; the canvas grew to unprecedented proportions and became the artist's field, or 'environment'. The act of painting became a formidable exercise in which gesture, in the service of self-expression, involved the entire body.

Art invaded walls and covered cities with 'graffiti'. Perhaps, ironically, chemicals derived mainly from the construction and automotive industries created a new chapter in painting. Polymer paint became the material of choice for large-scale abstractions. It could be applied as thickly as oil, dried fast and could be thinned, removed and used for creating an unlimited array of visual effects.

SPIRAL JETTY, ROBERT SMITHSON, 1970
This huge spiral of earth and rocks extends into Utah's Great Salt Lake. The jetty was submerged by a flood soon after it was built; the waters have risen and receded several times, revealing the earthwork salt crystals as they resurface. Smithson, who had a particular interest in science, helped to open up the field of cultural politics.

Starting with Robert Rauschenberg's *Combines* (1960), all manner of materials became potential ingredients for art and this notion has extended to satellites, and to the earth itself. While painting gave up its illusory third dimension, sculpture flourished with Calder's mobiles, Cesar's compressions, and installations of all kinds.

Roald Hoffmann, who received a Nobel Prize for Chemistry, wrote: 'The chemist chooses the molecule to be made and a distinct way to make it. . . . This is not so different from the artist who, albeit constrained by the physics of pigments and canvas, and shaped by his or her training, nevertheless creates what is new. The synthesis of molecules puts chemistry very close to the arts. We create objects that we then study, for others to appreciate.'

Ad Reinhardt (1913–67), who produced all-black paintings towards the end of his life, announced in the 1960s that painting was dead. Many then thought that he was right, and some still do. From dematerialized forms to physically aggressive works, art has since then taught us how to interpret industrial parts and materials as creation.

The end of painting?

Will painting be made obsolete by the current surge in scientific developments? This seems quite unlikely. New art forms derived from modern technologies are not competing with traditional media. On the contrary, they supplement the creator's choices.

In the past, artists have felt overtaken by emerging techniques. From camera obscura and photography to synthetic products and computers, they have systematically managed to turn perceived threats into opportunities. Artists nourish their passion using anything at hand, pieces of string or science . . . as long as it fuels their imagination.

The plethora of styles that has emerged in recent years gives us a full account of the painter's ability to respond to change. However, contemporary art also reflects a sense of confusion as we approach the next millennium with technologies capable of mass destruction.

While scientists tend to acknowledge that they do not necessarily possess the truth, artists in turn sometimes want to improve the world. Conceptual and contextual art has acquired a documentary dimension: some artists aim to reveal 'truths' in the way that scientists are expected to do.

Virtual Studio, Kriki, 1993
This interactive artwork invites the viewer to explore the 'painting' through a stereoscopic helmet, personalize his journey and keep a record (details above).
200 × 200 cm. Acrylic on canvas with laser CD. Photo P. Sebert

ENIGMA, ISIA LEVIANT
After a few moments of observation the colour rings apparently start to circle. Such works of art help to reveal how the brain interprets signals. Scientists can now localize specific brain cells which interpret colours, shapes and other signs used by abstract painters to emotional ends . . . fifty years ago.
On exhibit at the Palais de la Découverte, Paris

Ever since Impressionism, art naturally integrates multiple aspects of science. Nearly all recent styles – Constructivism, pop art, cybernetic art, and even space art – refer, one way or another, to technology.

But, after five centuries of uninterrupted triumph, painting seems to be short of breath. It is difficult to constantly re-invent the wheel for painters perpetually exposed to all the masterworks that were ever created.

Geniuses simply manipulating a paintbrush may become scarce, but were they not always? Some will continue to fascinate. Indeed, it is interesting to note that two of the most impressive twentieth-century artists, Francis Bacon and Lucian Freud, both representational painters remote from fashion, scrutinized humankind with a clinical, even a surgical, gaze.

Scientists study visually-induced emotions at the molecular level, but what makes us really appreciate a painting remains a mystery. The art historian, Ernst Gombrich, said: 'The biology of aesthetic pleasure has to take account of differences between civilizations and individuals [but] our reactions in front of works of art are simply too complex to be analysed scientifically.'

Although the science behind emotions is being slowly unravelled, much needs to be elucidated. One finding emerging in this field is that intellectual and emotional processes are strongly interrelated. Artists intuitively confirm this observation. Painting, the purest of art forms, has throughout history travelled parallel paths to science. Visual messages sent out by paintings defy what we understand of brain circuitry.

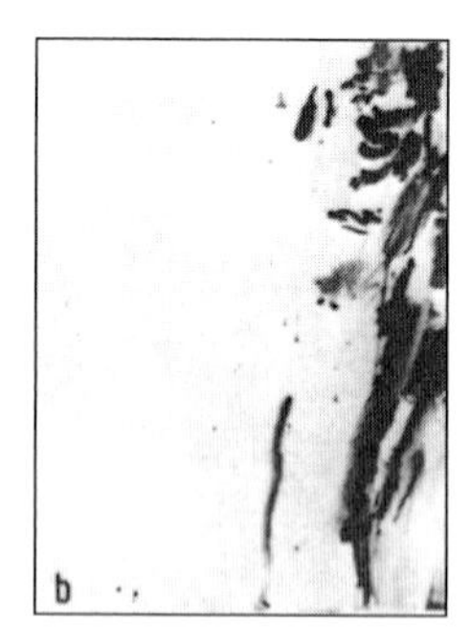

SELF-PORTRAITS BEFORE AND AFTER A STROKE, ANTON RÄDERSCHEIDT

The reference portrait above was painted two years before the artist suffered a brain injury. Researchers analysed the painter's drawing progress during healing. The left side of the canvas was, at first, systematically ignored by the painter; then his style became more expressive than his original manner – as if inhibition had been eliminated.

R. Jung. Psychiatrie der gegenwart, 1974

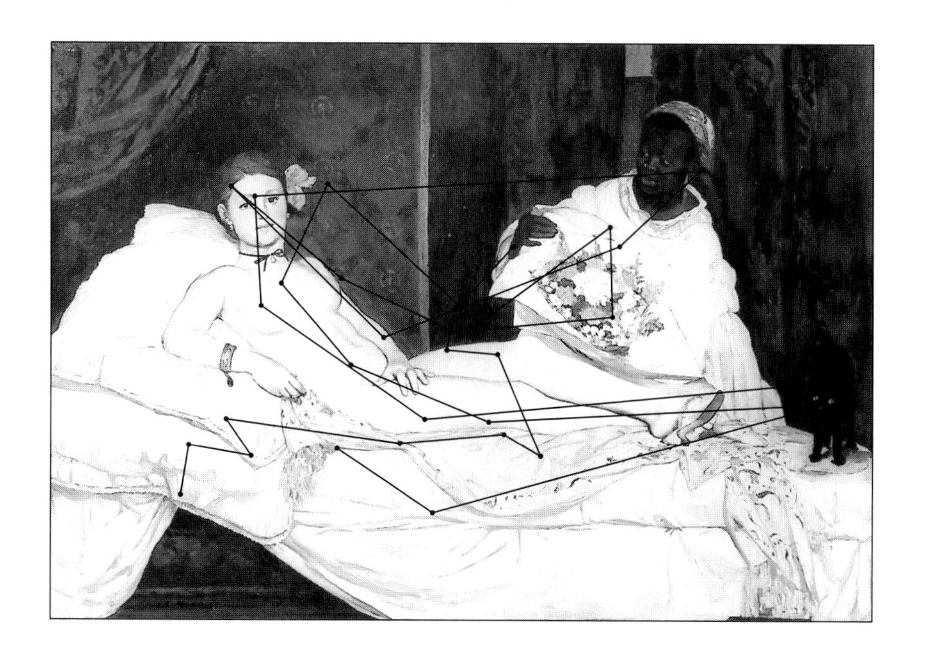

Scanpaths of a student's eye movements while looking at Manet's *Olympia*,
François Molnar
Such studies help scientists to unravel the mechanisms of aesthetic pleasure.
Musée d'Orsay. RMN. Courtesy of Vera Molnar

The language of graphic design

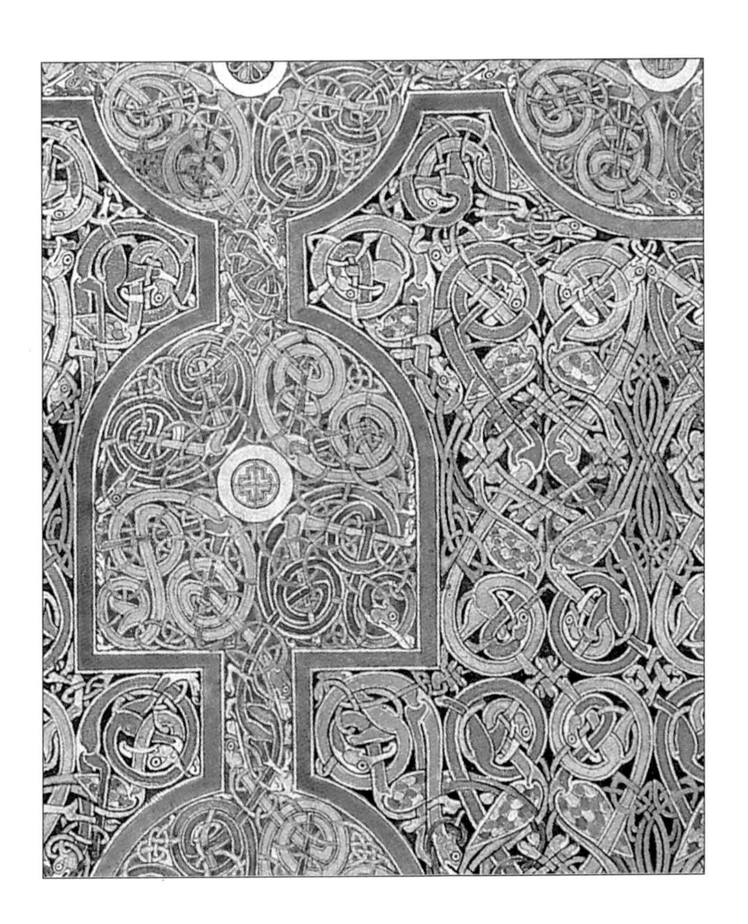

PAGE FROM THE *LINDISFARNE GOSPELS*, IRELAND, C. 700 A.D. (DETAIL)
This folio exhibits a breathtaking complexity. Geometric interlacing not unlike that found in Arab art, as well as wild pagan monsters, served as a vehicle to illustrate the Gospels. For all their apparent licence, these images were made according to rules and mathematical schemes. Organic and geometric shapes are kept strictly separate and, within each animal compartment, every line is connected to the animal's body.

Mss. Cotton. Nero.D.IV. folio. 26v. By permission of The British Library, London

Drawing and writing

Science has influenced most human activities through the emergence of novel techniques. Prehistory was defined by the Stone, Bronze and Iron Ages, referring to the presence of the relevant technologies. Likewise, history – the era of written communication – is classified by the making of clay tablets, parchment, typesetting and computing.

Today, the demand for visual communication is at an all-time high, accelerating the development of ever-new techniques. Our near future is perceived in cybernetic terms as television, video games, computers and other electronic devices have all become part of a single interactive network – blurring the traditional divisions of art and science.

POEM BY PAUL VERLAINE

In comparing the handwritten with the typed version, one cannot but think that with each new technological advance we benefit, learn and . . . lose a little.

Bibliothèque littéraire Jacques Doucet
Photo J.-L. Charmet

Il pleure dans mon cœur
Comme il pleut sur la ville.
Quelle est cette langueur
Qui pénètre mon cœur ?

Ô bruit doux de la pluie
Par terre et sur les toits !
Pour un cœur qui s'ennuie
Ô le chant de la pluie !

Il pleure sans raison
Dans ce cœur qui s'écœure.
Quoi ! nulle trahison ?
Ce deuil est sans raison.

C'est bien la pire peine
De ne savoir pourquoi,
Sans amour et sans haine,
Mon cœur a tant de peine !

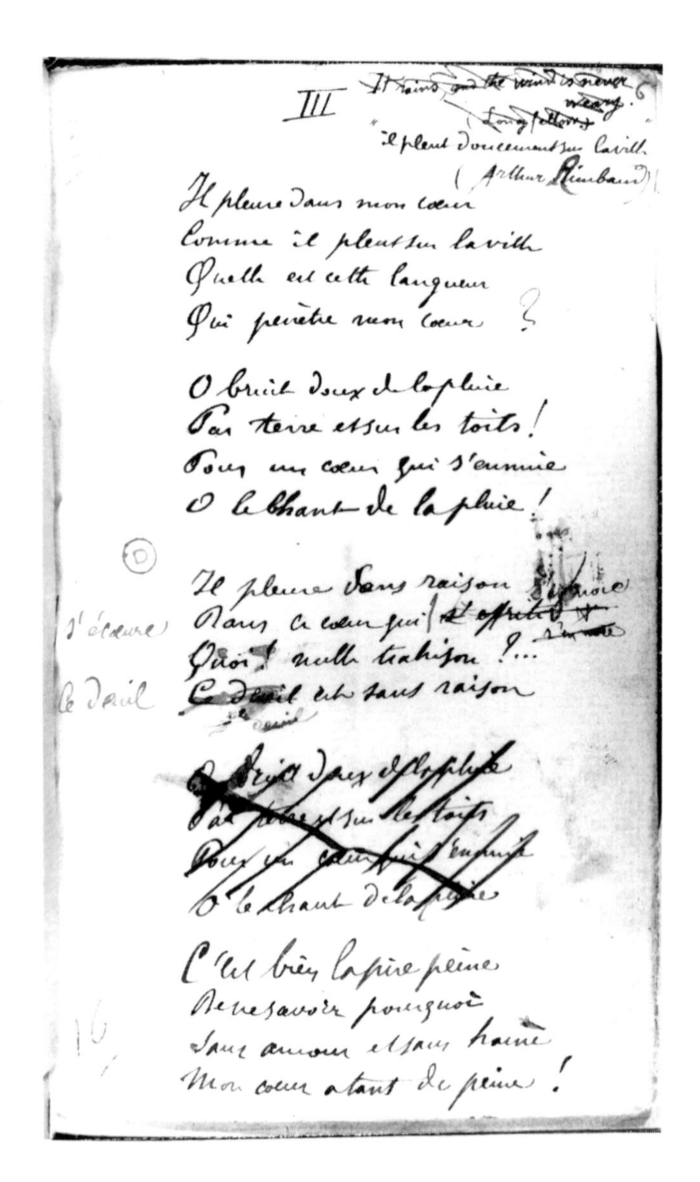

III

"Il pleut doucement sur la ville" (Arthur Rimbaud)

Il pleure dans mon cœur
Comme il pleut sur la ville
Quelle est cette langueur
Qui pénètre mon cœur ?

O bruit doux de la pluie
Par terre et sur les toits !
Pour un cœur qui s'ennuie
O le chant de la pluie !

Il pleure sans raison
Dans ce cœur qui
Quoi ! nulle trahison ?...
Ce deuil est sans raison

C'est bien la pire peine
De ne savoir pourquoi
Sans amour et sans haine
Mon cœur a tant de peine !

Prehistoric man apparently began painting, drawing and carving at an early date in different parts of the world. While colour is an emotional component, drawing and writing are more intellectual activities.

The oldest-known pictures had a ritualistic or utilitarian purpose, and hence what we view as the beginning of art should also be regarded as the dawn of visual communication.

'Primitive' art is often associated with abstraction, yet early representations were highly naturalistic; only much later did formal simplification occur. By the late Stone Age, pictographs – early writing with figurative or symbolic drawings – were so stylized that the step towards written characters seemed a relatively small one.

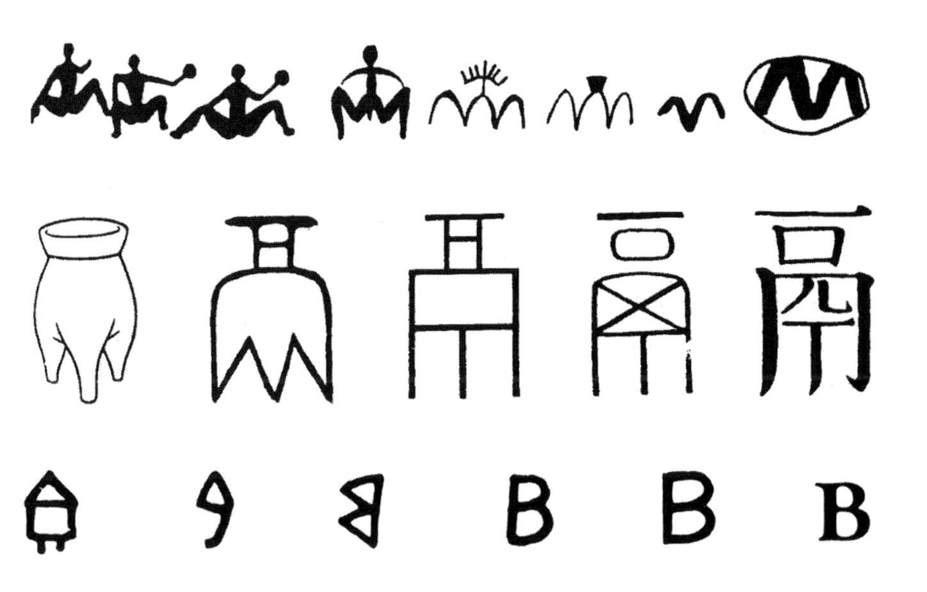

PREHISTORIC DRAWING ANALYSED BY ABBÉ BREUIL (ABOVE)
CHINESE ART ANALYSED BY SILCOCK (CENTRE)
DEVELOPMENT OF THE LATIN ALPHABET (BELOW)
Many forms of writing are derived from the gradual schematization of drawings. By combining pictograms it became possible to express ideas. Then, in some languages, ideograms were used to express the sound corresponding to the idea. Thousands of symbols could be summarized in a few phonetic symbols, and thereafter translated into alphabetical signs. The Chinese, on the other hand, chose to keep their numerous beautiful characters.

Writing was developed in Mesopotamia over 5,000 years ago and contributed enormously to civilization, as it preserved hard-earned knowledge and passed it on to future generations. At an early stage, grids contained the pictographs within spatial divisions; they were eventually turned on their side and written linearly in horizontal and vertical rows.

Pictographs were abstracted into a system of cuneiform (wedge-shaped) signs, which were inscribed into clay tablets. Speed increased when triangular-tipped styluses replaced the traditional pointed ones, as the new implement could be pushed instead of dragged through the clay.

The need for efficient record-keeping alone was sufficient to keep scribes employed in officiating over large-scale temples, eager to find techniques to facilitate their labour. Information was already an important business.

The library of King Assurbanipal (*c.* 668–25 B.C.) contained over 20,000 clay tablets inscribed not only with commercial facts and figures, but also texts on history, medicine, astronomy and astrology.

Over a 2,000-year period, several languages were written in cuneiform characters, which were gradually simplified.

CLAY TABLET WITH MEDICAL TEXT IN CUNEIFORM CHARACTERS, NIPPUR, IRAQ, *C.* 2100 B.C.
In Mesopotamia, clay was the prevalent writing material, although drawing shapes in wet clay is not an easy task. This is probably one of the reasons why a simple linear system was developed there.
Object 14221. neg # 58-55887.
University of Pennsylvania Museum, Philadelphia

TYSZKIEWICZ CYLINDER SEAL, OVERALL VIEW AND IMPRESSION OF SIDE AND BASE, HITTITE, FOURTEENTH CENTURY B.C.
This seal, which is both decorative and practical, was used for signature. Motifs ranged from figurative to totally abstract.

Hematite handle on cylinder of intaglio gem. # 98.706, height 5.8 cm, diameter 2.2 cm. Henry Lillie Pierce Fund. Courtesy Museum of Fine Arts, Boston.

The Egyptians, who were in contact with the Mesopotamians, decided to develop a different writing system based on hieroglyphs ('sacred sculptures' in Greek) – which embodied the beauty of their universe. It was a highly complex system based on a combination of different types of figurative, symbolic and phonetic signs.

Becoming a scribe required years of initiation. The highly respected guardians of knowledge were trained in land-surveying and accountancy, as well as in art and science. They kept their writing system for restricted usage and developed simplified forms for business and popular purposes.

Egyptians wrote on papyrus – made of moistened strips of thin slices of reed pith, superimposed at right angles and pressed – instead of on rigid tablets. Because the Egyptians wrote on papyrus scrolls, they perhaps produced the first documents that combined words and images.

Along today's Syrian-Lebanese coast, Phoenicians, around 1650 B.C., using the cuneiform system on papyrus, simplified it to twenty-two totally abstract phonetic signs. The alphabet is a series of visual symbols, each standing for an elementary

DETAIL OF A GREEK PAPYRUS MANUSCRIPT,
C. FOURTH CENTURY B.C.

Even though Greek, Latin and Carolinian alphabets have the same origin, each culture has left its own mark. Greek characters, like those used to transpose Homer's *Iliad* and *Odyssey*, are harmonious and rhythmic. The words *logos*, *harmonia* and *arythmos* have related meanings.

Ägyptisches Museum im Bodemuseum. Staatliche Museen Berlin. © bpk. ALS Colordia

sound. These signs soon spread throughout the ancient seafarers' world.

Who brought the alphabet to the West is not known for sure, but mythology points towards Cadmus of Miletus, also credited for having invented prose. The Greeks called the imported documents *byblos*, which was the name of their Lebanese export centre (*byblos* means 'book'). They adopted the simple and democratic alphabet, and developed its utility and beauty, changing five of the consonants into vowels as connecting sounds to make words.

In Ancient Greece, more than in any other civilization, beauty and knowledge went hand in hand. It is tempting to think that the Greeks, too, illustrated their manuscripts, but we have no proof of this. By the time of Alexander the Great (356–23 B.C.), the demand for information required an orchestrated, large-scale effort. He founded several libraries, of which the major one in the port of Alexandria housed up to half a million manuscripts. Anchoring ships were required to hand over documents for rapid copying. Each visitor had to lend them for that purpose and ships were searched to make sure nothing was withheld. A famous quarrel took place when Athenians lent manuscripts of Greek tragedies and received copies back, instead of the originals.

When Alexander's empire was split up after his death, his generals unwittingly spread civilization by broadly introducing the Greek alphabet. Languages such as Etruscan and Latin developed as a result, as did later Armenian and Cyrillic – after the Byzantine priest Cyril (ninth century) who devised an alphabet for the Slavs' language.

ROMAN CARVED INSCRIPTION ON TRAJAN'S COLUMN,
SECOND CENTURY A.D.

Roman lettering is sculptural and majestic. The artisan first calculated the number of letters and their size, making a scheme on papyrus, before drawing on stone and engraving the inscription. The Romans anticipated the shadows of the sculpted capitals, and incorporated these in their writing style (derived from *stylus*, the object used by the Romans for writing).

© The Board of Trustees of the Victoria & Albert Museum V&A Picture Library, London

Most ancient repositories of culture were lost; for example, of the 330 known Greek plays, only a few dozen have survived. Yet from Hellenistic territories the legacy of antique art and science spread east and west.

In Greek, the word *graphein* applied similarly to writing, drawing or painting. There also was an analogy between graph and glyph (sculpted image). As in ancient pictographic principles, most alphabets exhibit unity of idea and image.

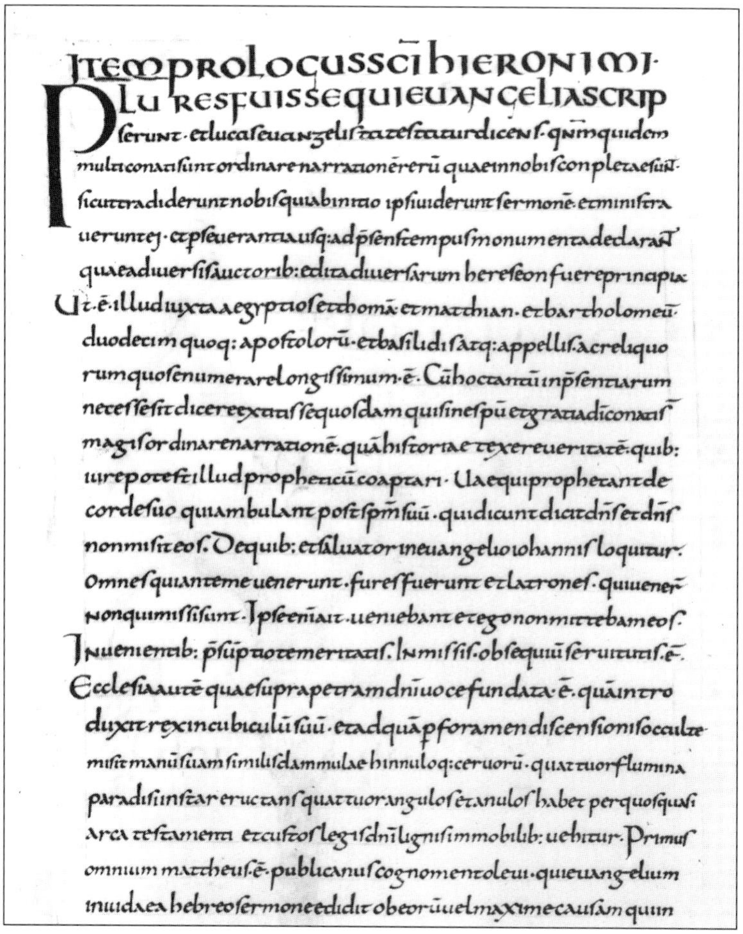

CAROLINE LOWER-CASE LETTERS IN THE ALCUIN BIBLE, NINTH CENTURY

The Romans developed a smaller, roundish writing style for daily use. Inspired by it, Carolinian writing, called 'Roman' by Charlemagne, looks even simpler and reflects an economical approach. According to legend, the great Charlemagne tried to learn how to write (hiding his notebooks under his pillow in order to practise at night) – but he never succeeded!

Add. MS 10546 Folio 351v.
By permission of The British Library, London

From parchment to books

The story goes that Ptolemy V of Alexandria and Eumenes II of Pergamon (Turkey, *c.* second century) were engaged in a fierce rivalry in the papyrus-scroll trade. As the Romans had embargoed Pergamon's shipments, its citizens proceeded to make parchment (which possibly takes its name from Pergamon) to fill the gap.

Made of animal skin, of which vellum is the finest form, parchment sheets were larger, smoother and more durable than papyrus. Even though parchment was more expensive, both sides of the page could be used, which considerably reduced storage space and also did away with the time-consuming necessity of rolling and unrolling the scrolls, which suffered considerable wear and tear in the process.

As the parchment writing surface became widely used, the change, around the fourth century, from scroll to codex ('book' in Latin) must have been a major stimulus to illustration. Parchment does not absorb colour, which makes pictures look brighter.

Calligraphy, and then illustrated texts, became a dominant art form which the Arabs turned into a tool for science – whose ancient form was rewritten and revised. Intellectual centres

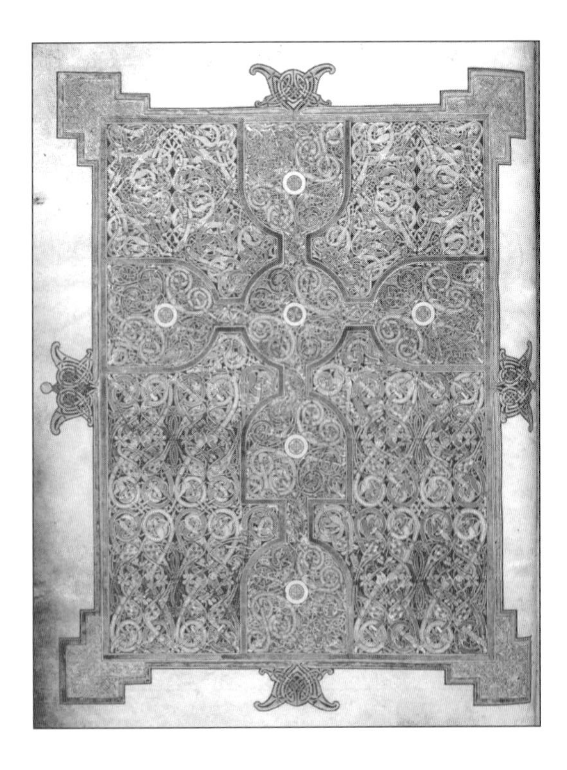

PAGE FROM THE *LINDISFARNE GOSPELS*, IRELAND, C. 700 A.D.
Patterns on parchment were first drawn on the back of the page. The key points of the design were pricked through the vellum to provide guidance to the painter. Measurement instruments such as the compass and the square were commonly used.

Carpet Page Mss. Cotton. Nero.D.IV. folio. 26v.
By permission of The British Library, London

developed across their territories: in Southern Spain, the Cordoba Library contained more books than the whole of France.

Later, Christianity became the primary fount of writing in the West. Manuscripts written in silver letters on purple vellum show that the art of miniature reached a high level early in the Middle Ages. Minium, a red-brown pigment made of a lead derivative mixed with egg, provided the name 'miniature'.

Religious manuscripts, whose pages were lavishly decorated with gold, silver and precious pigments, were produced in monasteries. Gold was ground or hammered on an adhesive background or applied as leaf, which produced the dazzling 'illuminated' effect. Monks often added touches of fantasy to their scholarly and decorative work. Monasteries in Ireland and England were prominent intellectual centres.

Manuscript production went into high gear during the reign of Charlemagne, who recruited Alcuin of York (735–804), a leading historian, as his personal advisor on education. Since few people could read or write, uneducated audiences had to be read to; readers and writers were distinct types of specialists.

Under Charlemagne's authority, the Roman alphabet was further developed: written texts were structured by punctuation marks and paragraphs, which simplified the transcription and the interpretation of manuscripts.

BENCHES AT TRINITY HALL, CAMBRIDGE UNIVERSITY LIBRARY, ENGLAND, *C.* 1590

Parchment led to the use of the goose quill. A quill was carefully selected, then softened by soaking for several hours. Once dry, it was sharpened and shaped as a function of the desired style. Because of its flexibility, the quill facilitated the development of new writing styles. Towards the end of the Middle Ages, lead pencils, paper and notebooks also contributed to the development of written communication. To meet the needs of a growing number of students, a system of lecterns and shelves was installed; books were secured by chains.

By permission of the Master and Fellows of Trinity Hall, Cambridge University

The most celebrated illuminated manuscript, *The Book of Hours*, was commissioned by the Duc de Berry (fifteenth century). He owned the largest private library in the Western world, containing over a hundred books, the production of which he personally oversaw. The miniature paintings in these volumes were remarkable for their detail and brilliant colours.

The production process was complex and expensive. Hundreds of sheepskins had to be prepared to furnish the high-quality vellum, to which precious metals and pigments were applied by teams of scribes and illustrators. The cost of a single book, with its elaborately carved cover inlaid with ivory and precious stones, could easily amount to the purchase price of a large estate, or a top-quality vineyard.

The art of copying antique documents, thus saving some of the greatest works, constituted the main artistic and scientific endeavour of the Middle Ages. But, with the invention of movable type just a few decades away, the thousand-year-old craft was doomed.

FEBRUARY, MINIATURE AND TEXT IN *THE BOOK OF HOURS,* POL LIMBOURG, FIFTEENTH CENTURY (OPPOSITE PAGE)

This picture provides the earliest representation of a snow landscape in the history of Western art. In the finest illuminated manuscripts, pictorial and written information reflect a high level of visual integration. Narrow Gothic writing characters were appreciated because many more could be fitted on a single page of costly parchment.

Musée Condé, Chantilly. Giraudon

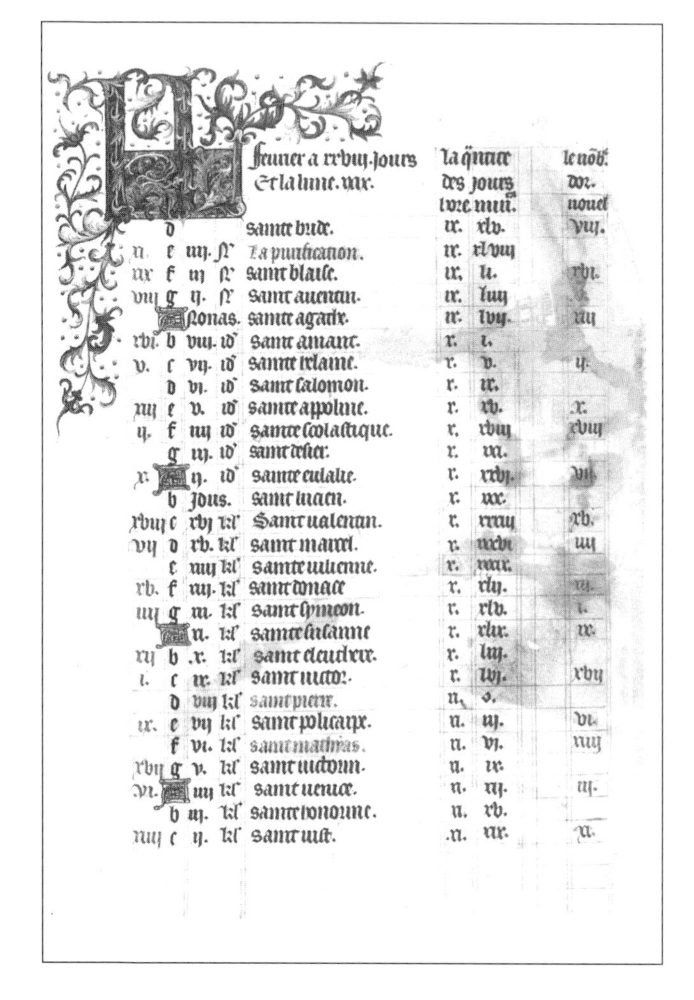

Death of a thousand-year-old art

According to legend, the ancient Chinese invented calligraphy (*c.* 1000 B.C.) by contemplating the clawmarks of birds and animal footprints. They wrote on wood or bamboo, cheap and readily available materials, or on silk which was costly to produce. Around 105 A.D., the eunuch Ts'ai Lun developed a process for fabricating paper and was deified as the god of paper-makers. In fact, paper had been known much earlier. As a lightweight material, it was perfectly adapted to hair or silk brush calligraphy. The finest scrolls were made by gluing sheets together and decorating them with materials such as jade.

Paper became used across Asia for various practical purposes. After winning the Battle of Samarkand (712), the Arabs took possession of Chinese paper-making secrets and jealously guarded them for another five hundred years. Paper was finally introduced to Europe in the twelfth century, where Ts'ai Lun's method continued to be used almost unchanged until the Industrial Revolution.

A simple relief-printing technique existed long ago in China. It was derived from an old custom of making inked rubbings from carved stones, and perhaps also from using Middle-Eastern engraved seals. Exactly how and when relief printing was invented is not known.

The oldest surviving relief-printed books (ninth century) attest to the incredible artistic quality attained by early

TRADITIONAL PAPER-MAKING TECHNIQUES FROM CHINA
The craftsman holds a tray that contains a mixture of disintegrated wood-pulp fibres. Various plants were used in the production of paper, much earlier in China than in the West. Anyone who has handled an ancient book made of linen knows how fresh it remains. Newer books with paper bleached by chlorine decay rapidly.
Ontario Science Center

Chinese printers. By this technique they made encyclopedic works, sometimes 100,000 pages long.

On pilgrimage routes there was a great demand for religious pictures, which were printed from carved slabs of stone or wood, and from bronze plates. The demand for playing-cards was high, too. (As in Ancient China, numerical sequences and royal figures feature in today's cards.)

The earliest characters of movable type were developed by the alchemist Pi Sheng (*c.* 1000 A.D.). Chinese writing contains more than 50,000 characters which are neither alphabetical nor syllabic, but logographic: each sign represents an entire word and there is no relationship between the written and spoken language. In addition, a single sign may carry multiple interpretations, depending on how it is pronounced. All this made typesetting very impractical.

More importantly, the Chinese placed a premium on calligraphy, in which breathing and concentration, body and physical language, are interwoven. Calligraphy was an art form considered superior even to painting, as well as a science whose mastery commanded the highest respect. Thus there was no profound reason to develop printing.

FUNERARY COUCH STONE, NORTHERN CH'I DYNASTY, 550–77 A.D.

The Chinese developed relief printing by taking high-quality inked rubbings from carved stones.

Dark grey limestone, 47 x 112 cm.
Gift of Denman Waldo Ross and G. M. Lane. Courtesy Museum of Fine Arts, Boston.

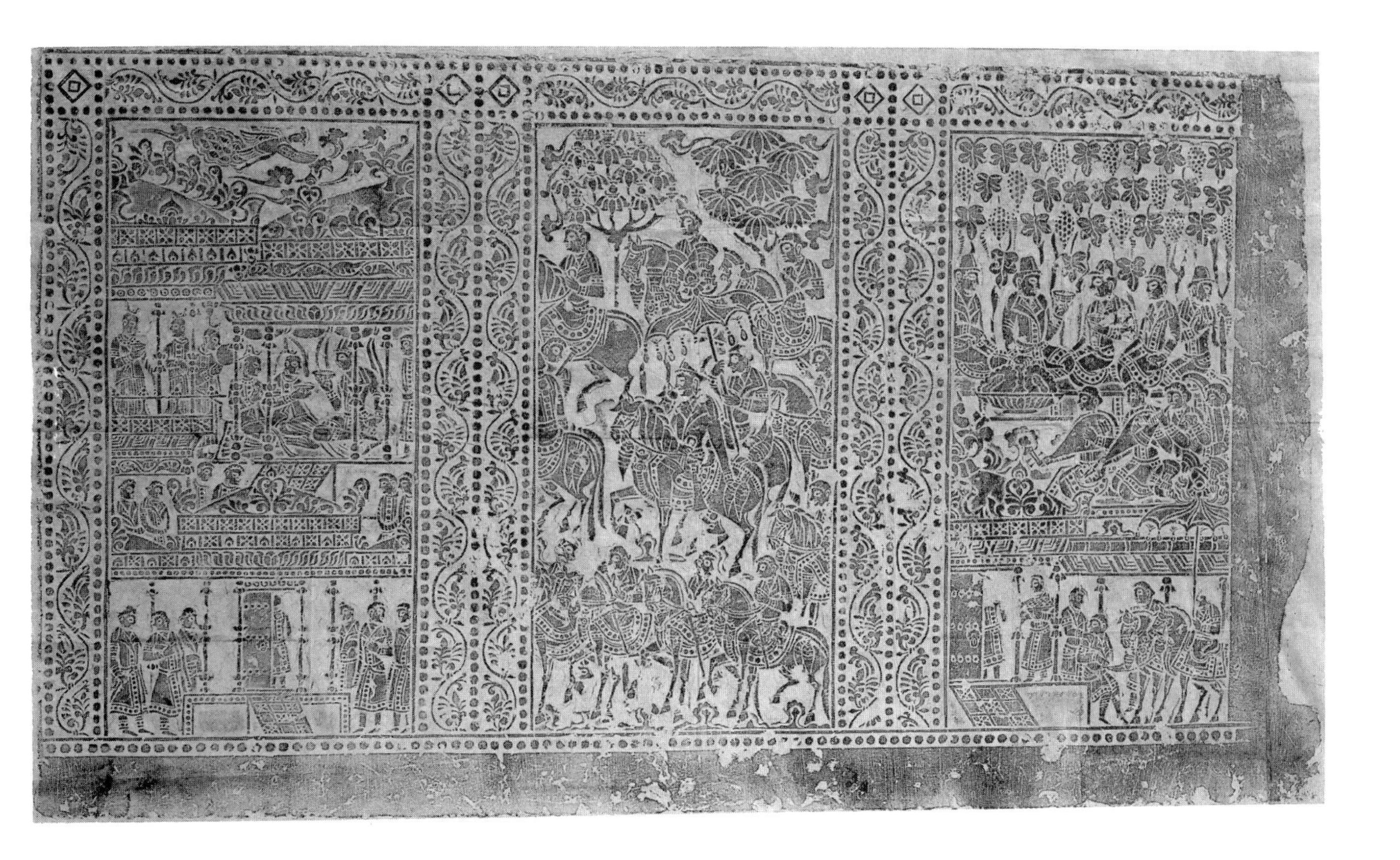

corde: qm̄ ipi deum videbūt. Beati pa-
cifici: qm̄ filij dei vocabūtur. Beati q̄
ꝑsecutionē patiūtur ꝓpter iusticiam:
qm̄ ipsoꝝ est regnū celoꝝ. Beati estis
cū maledixerint vobis· et ꝑsecuti vos
fuerint et dixerūt· om̄e malū aduersū
vos mentientes ꝓpter me. Gaudete ⁊
exultate: qm̄ merces vr̄a copiosa est ī
celis. Sic enī ꝑsecuti sunt ꝓphetas qui
fuerunt ante vos. Vos estis sal terre.
Qd̄ si sal euanuerit: in quo salietur?
Ad nichilū valet vltra· nisi ut mitta-
tur foras: et cōculcetᷣ ab hominibus.
Vos estis lux mūdi. Non potest ciuitas abscondi supra mōtem posita: ne-
q; accendūt lucernā· et ponūt eam sub
modio sed sup candelabrū: ut luceat
om̄ibus qui in domo sunt. Sic luceat
lux vr̄a corā hominibus: ut videant
opera vr̄a bona: et glorificent patrē
vestrū q̄ in celis est. Nolite putare qm̄
veni soluere legem aut ꝓphetas. Nō
veni soluere: sed adimplere. Amen
quippe dico vobis: donec trāseat celū
et terra· iota vnū aut vnus apex nō
ꝑteribit a lege: donec om̄ia fiant. Qui
ergo soluerit vnum de mandatis istis
minimis· ⁊ docuerit sic homines mi-
nimus vocabitur· in regno celoꝝ: qui
aūt fecerit et docuerit: hic magnꝰ vo-
cabitᷣ in regno celoꝝ. Dico aūt vobis:
quia nisi abundauerit iusticia vestra
plusq̄ȝ scribaꝝ et phariseoꝝ: nō intra-
bitis in regnū celoꝝ. Audistis quia
dictū est antiquis nō occides: qui aūt
occiderit reus erit iudicio. Ego aūt dico vobis: q̄a om̄is q̄ irascitᷣ fratri suo
reus erit iudicio. Qui aūt dixerit fratri
suo racha: reus erit cōcilio. Qui aūt dixerit fatue: reus erit gehenne ignis. Si
ergo offers munꝰ tuū ad altare ⁊ ibi
recordatꝰ fueris quia frater tuꝰ habet
aliquid aduersū te: relinque ibi mu-
nus tuū ante altare: et vade prius re-
conciliari fratri tuo: et tūc veniens of-
feres munꝰ tuū. Esto cōsentiēs aduer-
sario tuo cito dū es in via cū eo: ne
forte tradat te aduersariꝰ iudici: et iu-
dex tradat te ministro· et in carcerem
mittaris. Amen dico tibi: nō exies in-
de donec reddas nouissimū q̄drantem.
Audistis quia dictū est antiquis: nō
mechaberis. Ego aūt dico vobis: q̄a
om̄is qui viderit mulierem ad concu-
piscendū eam: iam mechatꝰ est eā in
corde suo. Qd̄ si oculus tuus dexter
scandalisat te: erue eum et ꝓice abs te.
Expedit enī tibi ut pereat vnū mēbro-
rum tuoꝝ: q̄ȝ totū corpꝰ tuū mittatᷣ in
iehennā ignis. Et si dextra manꝰ tua
scandalisat te: abscide eā et ꝓice abs te.
Expedit enī tibi ut pereat vnū mēbro-
rum tuoꝝ: q̄ȝ totū corpus tuū eat in iehennā. Dictū est aūt quicūq; dimiserit vxorem suā det ei libellum repudij.
Ego aūt dico vobis: quia om̄is q̄
dimiserit vxorē suā· excepta fornicatio-
nis causa· facit eā mechari: ⁊ q̄ dimis-
sam duxerit adulterat. Iterū audistis
q̄a dictū ē antiquis nō ꝑiurabis: red-
des aūt dn̄o iuramēta tua. Ego aūt
dico vobis nō iurare om̄ino: neq; ꝑ ce-
lum· q̄a tronꝰ dei ē: neq; ꝑ terrā· q̄a sca-
bellū ē pedū eiꝰ: neq; ꝑ iherosolimā· q̄a
ciuitas ē magni regis. Neq; ꝑ caput
tuū iurauerīs: quia non potes vnū
capillū albū facē aut nigrū. Sit aūt
s̄mo vester: est ē· nō non: qd̄ aūt hijs
abundantiꝰ ē a malo ē. Audistis q̄a
dictū ē: oculū pro oculo· dentem ꝓ den-
te. Ego autem dico vobis nō resistere
malo: sed si quis te ꝑcusserit in dextrā
maxillā tuā ꝓbe illi ⁊ alterā: et ei qui
vult tecum in iudicio contendere et

PAGE FROM THE *GUTENBERG BIBLE*, 1450–55

Gutenberg was the father of Western printing. For two decades, he experimented to determine a suitable metal alloy that would flow easily when heated and make reproducible-quality characters, various letter designs and punctuation marks, the right black and coloured inks – probably based on Flemish painters' experiences – the right kind of paper and, most importantly, the optimal 'screw press' for making impressions in large numbers, daily. When the first typeset book appeared, the *Gutenberg Bible,* its visual clarity and aesthetic qualities heralded typography as a superior art form.

Paper and various concepts relating to movable type were progressively introduced in the West, where they found a knowledge-hungry audience with a growing need in business administration. The development of typesetting, that is, printing with independent, reusable standardized characters was, after writing, the most important step in the advancement of civilization.

It is generally accepted that Johannes Gutenberg (1394–1468) brought the different technologies together. From his father who worked at the Mainz mint, he learned all the tricks of working metal. A successful metal- and gem-cutter, he spent decades in developing the printing press.

After nearly two decades of Gutenberg's research, Johannes Fust, who had provided financial backing, suddenly sued for not having received interest on the loans. Fust created the largest worldwide printing firm, having taken all of Gutenberg's assets, including the famous Bible. It is said that the legend of Faust, in which the protagonist exchanged his soul for power, had its source in Gutenberg's misfortune.

Gutenberg's printing technique was no mere collection of technical recipes. The physical presentation and visual impact of texts were prime concerns from the very beginning, which also explains why illuminated manuscripts ceased to be produced within a few decades.

Over the same period, printing had become established in the West. By 1480, presses existed in more than twenty northern towns and twice as many appeared in the south. In Venice, Aldus Manutius (1449–1515) translated and printed nearly all the antique works available, the content of which was saved for posterity.

While 'How to do it' manuals appeared in every possible field, Venetians complained that there were already too many books, thus the portable format was adopted.

The notion of authorship grew in importance, as did indexing and filing as collections rapidly developed. Sequential ordering of pictures, words and ideas, provided in a reproducible and verifiable form, were the conceptual pillars of Western progress.

Birth of modern illustration

Drawing changed radically when paper was introduced. Leonardo da Vinci (1452–1519,) in the south, and Albrecht Dürer (1471–1528), in the north, established the new way of drawing as a common system, integral to the graphic arts, architecture and science alike.

Leonardo used perspective theory and practice, trusting only direct observation and his own experience. The object of his life was *saper videre,* or to know how to see. 'He who loses his sight loses his view of the universe. The eye has measured the distance and the size of the stars, [it] has created architecture and perspective, and lastly, the divine art of painting.'

He distinguished three aspects of perspective: linearity, colour and blurring. The first deals with the apparent diminution of objects as they recede from the viewer. The second relates to changes in colour as a result of recession from the

FRONTAL VIEW OF A LION,
VILLARD DE HONNECOURT, C. 1235
Drawing in the Middle Ages was related to the flat and static architecture of stained-glass windows. Villard de Honnecourt, a widely-travelled medieval builder who illustrated his projects, said 'the art of geometry commands and teaches'. In this picture of a lion, he laid down noticeable circles, one for the face and another larger one for the body.
cod. fr. 19093. Photo Bibliothèque Nationale de France

eye, and the third aspect accounts for the blurring effect, as receding objects appear less distinct.

Leonardo gave a detailed description of a pyramidal theory of perspective: 'Painting is based upon perspective which is nothing else than a thorough knowledge of the function of the eye. And this function simply consists in receiving in a pyramid the forms and colours of all objects placed before it. Therefore, if you extend the lines from the edges of each body as they converge you will bring them to a single point, and necessarily the formed lines must form a pyramid.'

He invented a new genre, modern illustration, which would be a determining factor in documenting and structuring the fast-growing body of information; while he analysed motion he developed a cinematic drawing style. Even though he experimented widely, and seemingly endlessly, Leonardo never wrote a formal treatise. Instead, he left hundreds of notes and sketches – in mirror-writing since he was left-handed – many of which were found three hundred years later. They thus had a limited impact on the course of science.

Leonardo immensely enriched the fields of ballistics and engineering, and conducted pioneering work in anatomy that covered embryology and physiology, as well as the study of expression and movement. Few of his inventions found practical application and only a small number of paintings remains: roughly fifteen can be attributed with certainty, and some of them are in a lamentable state of conservation.

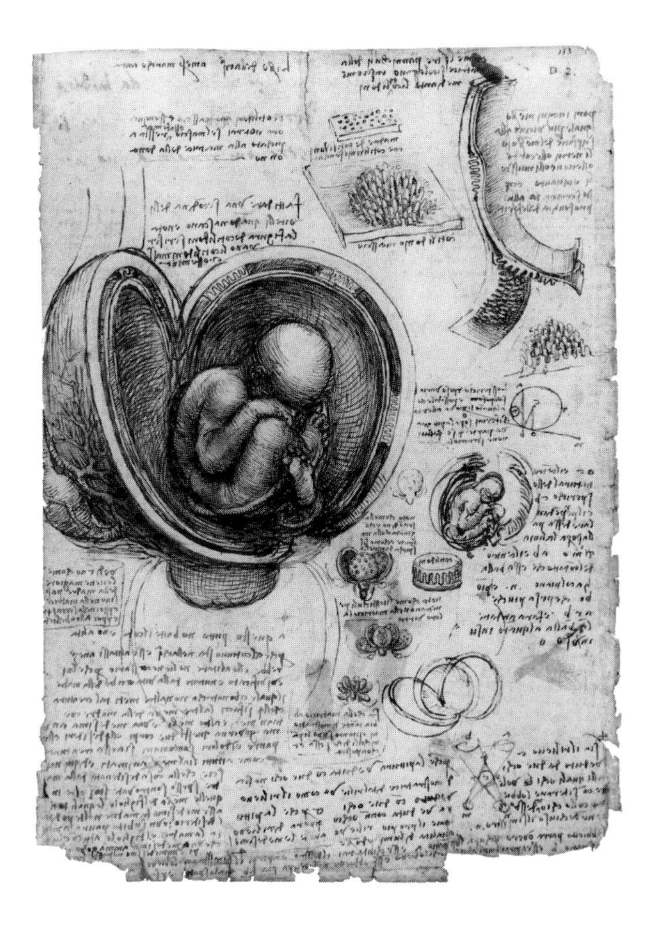

THE INFANT AND WOMB, DETAIL OF PEN DRAWING, LEONARDO DA VINCI, SIXTEENTH CENTURY

The victims of the plague in the Middle Ages left behind tons of garments, which were used for manufacturing paper. Rapidly executed design such as this was directly linked to the availability of paper. Drawing evolved spectacularly during the Renaissance. Anatomical sketches were made in conjunction with the dissection of corpses. Figures were first conceived as nude drawings. Then, bodies were sculpted by light and clothed by colour.

CHAMPFLEURY ALPHABET, GEOFFROY TORY, SIXTEENTH CENTURY

Typography became an art, as indicated by the above characters. Using positions of the human body, Tory designed all the alphabet characters by means of a central zero, representing the sun, whose twenty-three rays correspond to: the nine Muses, the seven Liberal Arts, the four Cardinal Points and the three Graces.

Photo Bibliothèque Nationale de France

His visionary scientific notes fascinate scholars, but his lasting achievement was, however, his art. As a painter, sculptor, architect, engineer, physicist, biologist and philosopher, Leonardo was invaluable to Western culture.

The first to make an organized attempt at transposing the Renaissance visual ideal to the North was the German painter and engraver Dürer, who concentrated on realistic representation. He was convinced that this could only be done by injecting scientific theory into art.

Dürer was highly regarded by Kepler and Galileo, having published the first German geometry manual. He said: 'A painter [reading this book] will not only get a good start, but will reach better understanding by daily practice.' Dürer considered that his theories had to be of use to draughtsmen and painters, but also to stone-, wood-, and metal-carvers, and generally to all those who used the art of measuring.

Graphic art created in this context had a three-dimensional, sculptural quality. Nurenberg, Dürer's home, was central Europe's most prosperous city. It was a printing capital and major producer of navigational maps and instruments. A goldsmith's son, Dürer had extensive experience with such devices, which were well-suited for drawing.

He was famous as an oil painter and pioneered new styles of portraits and nudes. As one of the earliest Renaissance watercolourists, the handling of light in his landscapes and nature studies looks astonishingly modern.

MAN DRAWING A LUTE WITH THE HELP OF A MECHANICAL DEVICE, WOODCUT, ALBRECHT DÜRER, SIXTEENTH CENTURY

Dürer suggested that drawing faults could be corrected by training according to defined methods. He proposed various mechanical aids such as the 'Through-The-Looking-Grid', in which the eye of the artist is guided by a frame divided into squares, placed between the artist and the model.

Spencer Collection. Astor, Lenox and Tilden Foundations. The New York Public Library

Dürer is mainly remembered for his engravings, transforming a craft into an authentic art form. Woodblocks were considered capable of transferring only simple images and were thus used mainly in the production of cheap prints. The multi-talented Dürer, however, succeeded in producing masterpieces from woodblocks.

Illustrators and printers collaborated with religious zeal. As a publisher, Dürer was a successful mass communicator; for him there was no hiatus between fine art and applied, or commercial art. German printing centres, thanks to the wide dissemination of their prints, had a direct impact on the Reformation. From cities such as Nuremberg and Wittenberg (Luther's headquarters), printing spread throughout Europe.

MELENCOLIA I, METAL ENGRAVING,
ALBRECHT DÜRER, 1514
In this symbolic image, the star pentagram stands for the Brotherhood of Pythagoras: its diagonals intersect each other according to the Golden Section. Other symbols can be identified in the sphere, scale and compass.
14" × 18". 59.805. Gift of Miss Ellen T. Bullard.
Courtesy Museum of Fine Arts, Boston.

Venice, where the trade routes crossed, led the way in book design. Numerous typographic innovations – roman and italic characters, printed pagination, lavishly ornamented title pages and sophisticated layout – made this city a capital of elegant book printing.

The latest developments spread from Italy, especially to France, and fostered a golden age. Punchcutters were in such demand that they gained independence from the printing companies, marking the division of publishers and printers. New typefaces, selected for their visual clarity, accompanied the introduction of the accent, the cedilla and the apostrophe, which significantly modified the structure of language.

With the persecution of the Huguenots, many of the French elite fled to England, Switzerland and the Netherlands, and took their professional skills with them. Printing empires, such as that of the Elseviers in the Netherlands, developed their activities around that time. The French printer, Christophe Plantin (*c.* 1520–89), ended up in Antwerp, where he created the largest publishing house of the time. He was acknowledged for his superior copper-engraving technique.

TITLE PAGE FOR *DE NATURA STIRPIUM LIBRI TRES*, SIMON DE COLINES, 1536

As book covers tended to be individually commissioned, owners awaiting the final embossed leather product could enjoy the temporary, elaborately decorated cover, which evolved into the title page. Printing was viewed as 'culturally strategic'. Louis XIV would call it the 'artillery of the intellect'. He appointed a committee of specialists headed by a mathematician to develop new writing characters.

Photo Bibliothèque Nationale de France

Metal is more durable than wood and is more resistant to the printing process, thus yielding a far greater number of impressions. More importantly, metal plates can be worked in much finer detail than woodblocks, thereby allowing artists to personalize their work. As soon as metal plates became available, engravers began to represent delicate materials, such as fur and feathers, and to suggest their specific textures.

Printmakers were often adventurous artists, as printing requires visual organization and a special capability for orchestrating various steps. It belongs to a vast research continuum: the development of yet another new technique, etching, would soon impress visual artists even more.

Etching may derive from a process developed in the fourteenth century for decorating weapons and repairing rusting shields. It involves drawing with a needle on a copper plate covered with a waxy film. When the drawing is finished the plate is bathed in acid, which bites into those places where the protective wax has been removed.

This chemical reaction thus substitutes for the exacting physical labour of scoring wood or engraving copper, whose material resistence made fine results difficult to achieve. Etching gave an expanded palette of tonal gradations and was adopted by the most creative artists.

Christ with the Sick around Him, Receiving Little Children, from *The Hundred Guilder* Print, Rembrandt, c. 1649

Rembrandt was a passionate experimenter who reworked his plates constantly, combining abrupt slashing of the copper with drypoint (scratching with a fine needle) and etching. He left several hundred printed masterpieces.

1943.287 28.2 × 39.5 cm.

Cincinatti Art Museum, Bequest of Herbert Greer French

Classification and conceptualization

Engraving, the standard method for illustration, generated a 'visual revolution' comparable with that created by television. Illustrated books proliferated in all domains – scientific, artistic, religious, musical and literary.

This welter of knowledge suddenly available to the public challenged philosophers such as the Frenchman, René Descartes (1596–1650) and John Locke (1632–1704) in England, to organize and interpret it. Baruch Spinoza (1632–77), a Dutch philosopher (who polished optic lenses to make a living) was another participant in the construction of Western values with his book on *Ethics Based on Geometric Order.*

The Almighty was the fount of all creation, but his omnipotence would soon be challenged by rational philosophers, who saw man and his environment in scientific terms, within the confines of mathematical laws. The 'Enlightenment', as seen by Immanuel Kant (1724–1804), a philosopher and physicist, was a broad reconciliation of English empiricism with French rationalism. Progress now seemed to be the essence of all things, including the chain of life.

While new branches of knowledge were in the making, imposing order on knowledge was also the mission of the French Count, Georges de Buffon (1707–88), and the Swede, Carolus Linneus (1707–78), who independently classified thousands of types of animals and plants, brought to them

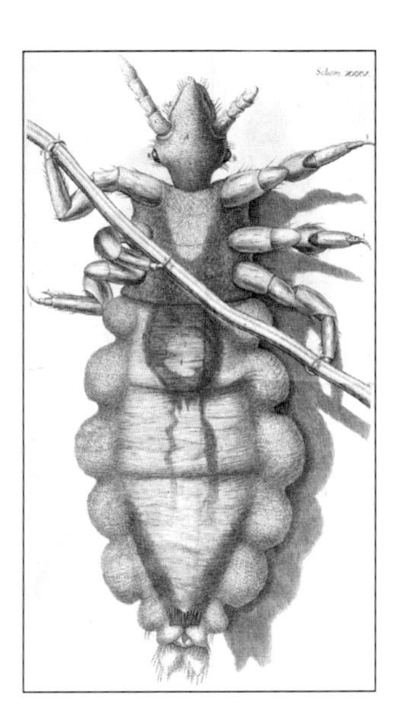

FLEA CLINGING TO A HAIR, ROBERT HOOKE, 1665
Hooke (1653–1703), London's Leonardo, was a microscopist and microscope designer. His *Micrographia* is a textbook on a variety of subjects including light and colour, combustion and respiration. Hooke coined the word 'cell', noticing that the inner structure of cork looked like cells in monasteries. The fundamental aspect of this observation became clear in the next century when it was discovered that cells are the basic unit of all living forms.
Science Museum, London.
(Source) Science & Society Picture Library

Sinesische Triümpfbogen, deren eine Menge in den grossen Stätten züsehen.
Arc Triomphal Chinois. On en voit une qvantité dans les grandes Villes

CHINESE TRIUMPHAL ARCH IN THE *OUTLINE FOR A HISTORY OF ARCHITECTURE*, 1721
JOHANN BERNHARDT FISCHER VON EHRLACH
The first modern book on architecture contributed to the dissemination of styles across borders and thus, to the development of this art form.
Österreichische NationalBibliothek, Vienna.
Photo Bildarchiv, ÖNB

from all over the world. They developed the disciplines of zoology and botany.

Buffon, who trained in mathematics and physics, ambitiously attempted to describe all living creatures in his 36-volume illustrated encyclopedia, *Histoire Naturelle.* He speculated that the earth was about 75,000 years older than the Bible suggested and even proposed a rudimentary theory of evolution that scandalized readers, a century ahead of Darwin.

Linneus, on the other hand, classified plants, by means of a simple system based on external features, still used in the twentieth century: a binomial identifying the genus (the group to which the plant belongs) and the species (a qualifying adjective) – like a surname and a name on which everyone could agree.

Linneus thus eliminated the traditional formula of cumbersome description which hampered the establishment of

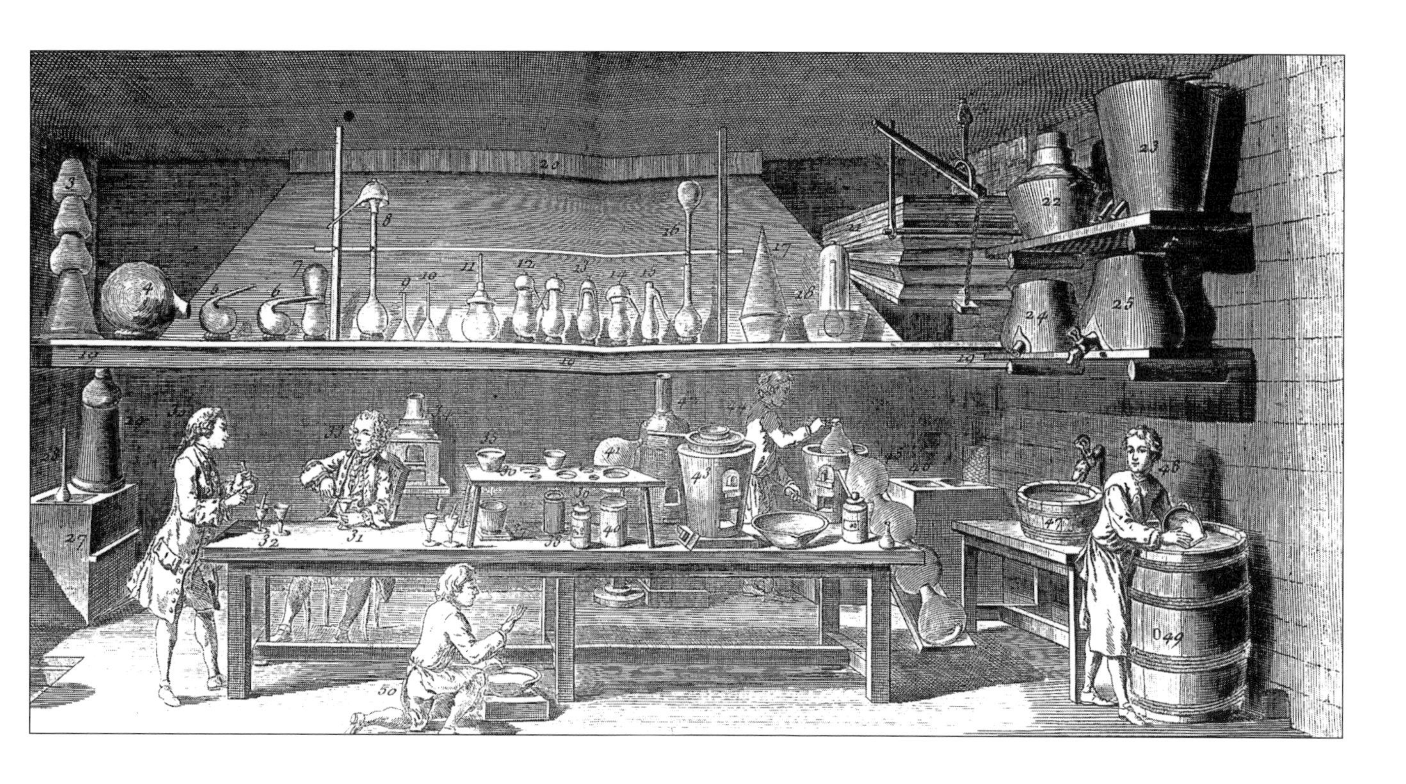

THE CHEMIST'S WORKROOM IN DIDEROT'S *ENCYCLOPÉDIE,* 1752
The first encyclopedia appeared in the United Kingdom. The French *Encyclopédie* consists of seventeen volumes of text and eleven volumes of illustrations. Jean d'Alembert and Denis Diderot were the brains behind this gigantic enterprise – Jean-Jacques Rousseau wrote the chapter on music. This work marked a turning point in comparison and critique.

systematic studies. He provided a framework for biology, in much the same way that Newton had provided one for physics in the previous century.

Since neither Descartes' nor Newton's systems provided all the answers, the next step was to try to order knowledge alphabetically. This is what Denis Diderot (1713–84), a French mathematician and philosopher, worked on for nearly thirty years with his colleagues, compiling the *Dictionnaire Raisonné des Sciences, des Arts et des Métiers,* called the *Encyclopédie.*

Diderot stated that the aim of the work was 'to bring together all the knowledge scattered over the surface of the earth, and thus to build up a general system of thought, so that the work of past ages would not be useless, and our descendants becoming more instructed should become more virtuous and happier'. The exhilarating exercise of democratizing knowledge eventually led to the belief that, if widely disseminated, it could provide a solution to all the problems of mankind.

In the lavishly illustrated works of Buffon, Linneus or Diderot, vast amounts of information were presented simply, coherently and artistically. The eighteenth-century illustrators paved the way for Darwin and his theory of evolution of species, implying natural selection instead of a pre-determined

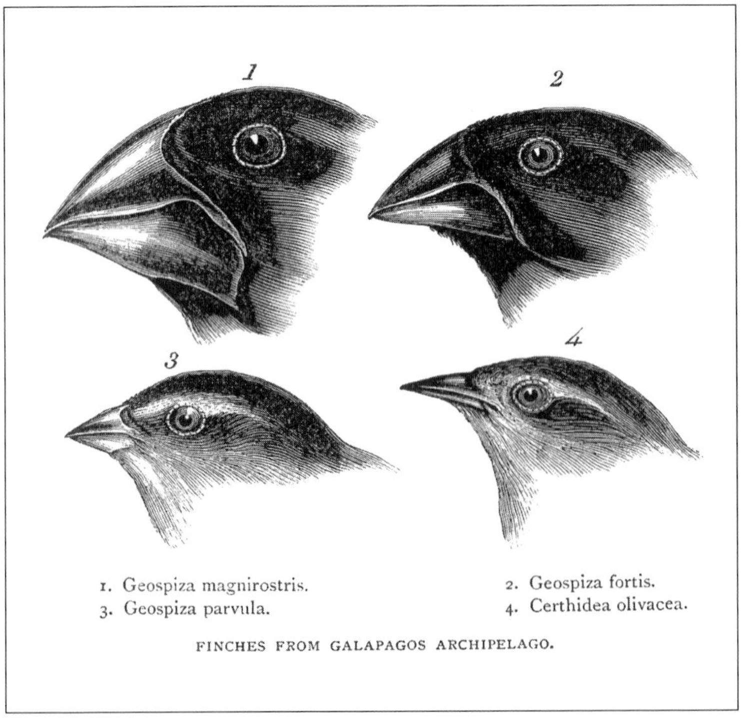

1 2 3 4

1. Geospiza magnirostris. 2. Geospiza fortis.
3. Geospiza parvula. 4. Certhidea olivacea.

FINCHES FROM GALAPAGOS ARCHIPELAGO.

GALAPAGOS ISLANDS FINCHES,
CHARLES DARWIN'S JOURNAL
Darwin carefully analysed drawings, such as these, after his voyage. It is reported that John Gould, an artist and ornithologist, focused his friend Darwin's attention on the link between birds' physical features and their ability to adapt to particular environments.
History & Special Collections Division. Louise M. Darling Biomedical Library, UCLA

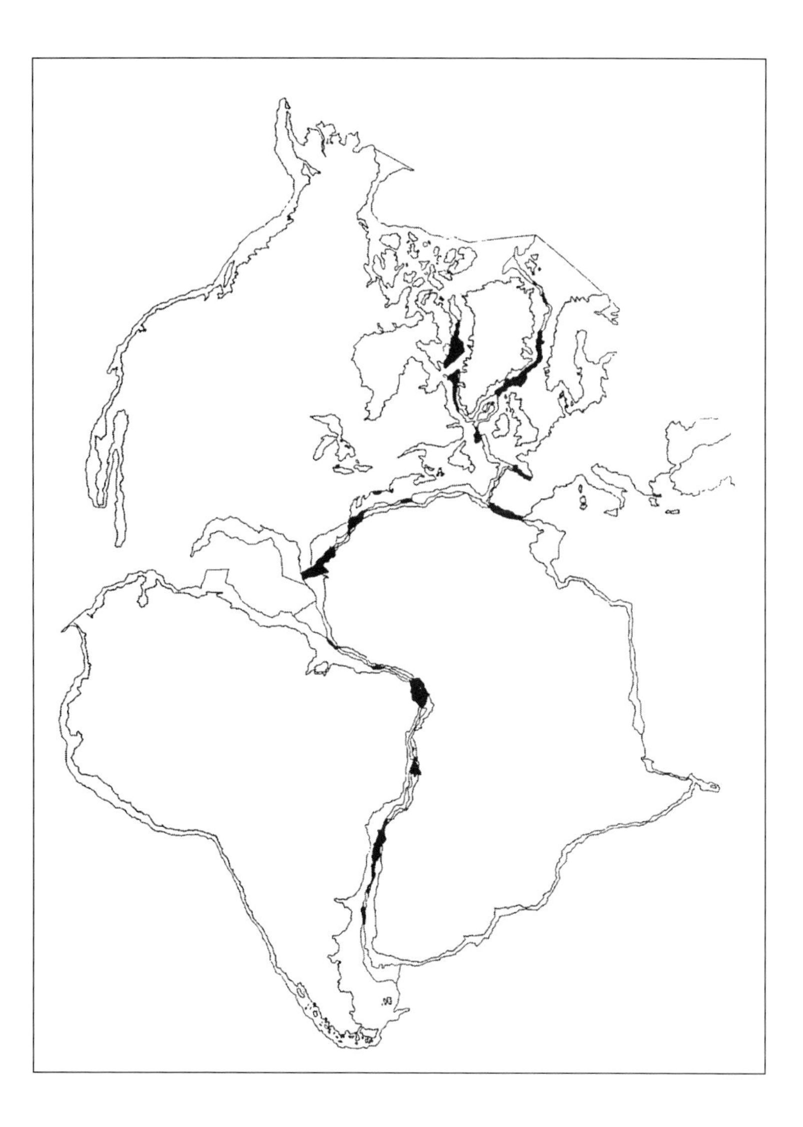

ORIGINS OF CONTINENTS AND OCEANS,
ALFRED WEGENER, 1915
While studying maps, Wegener noticed that the east coast of South America and the west coast of Africa could be fitted together like pieces of a jigsaw puzzle. The contours of the land masses gave him the idea that the two continents might have been joined at some time in a distant past, and that all continents had drifted away from a single one called Pangaea. This idea was recently validated by an international team of geophysicists, oceanographers, palaeontologists and biologists.
GB60.W4. Library of Congress, Washington, D.C.

scheme. Darwin was very lucid about the revolutionary aspects of his work, and was confident that with additional proof, the next generation would understand it.

Throughout history, illustrators – by accurately recording their observations, making models and sketches, designing easy-to-follow diagrams and contributing methodologies – were instrumental in the development of classification and comparative analysis, undermining age-old concepts of right and wrong.

With the cult of knowledge, researchers in all fields could hope to shape culture, instead of merely submit to what was dictated. Philosophers paved the way for the human sciences, whose laws were a counterpart to the mechanical principles describing natural phenomena. But the crowning triumph of all this activity was one of technical progress in visual communication.

The Industrial Revolution ushered in the age of mass communication. William Hogarth (1697–1764), a prolific caricaturist and clock engraver, demanded precision: he was the one who pressed the British Parliament to protect artists against the unauthorized use of their images in printed reproduction. Thus the copyright law was born.

As time went by, printing presses, which had scarcely been changed since Gutenberg's time, were modernized into high-speed machines, built from sturdy cast iron and powered by steam. This allowed for several thousands prints per day. Then even more sophisticated machines appeared: cylinder plates, inked rollers, new paper-producing machines and linotypes (from 'line o' type'). The latter composing machines dramatically reduced the price of newspapers and were not improved upon until the mid-twentieth century, when photocomposition was developed. As literacy increased, a tidal wave of eclectic typographic designs flooded the pages of newspapers and books.

TO THE INHABITANTS OF THE PARISH OF

CLERKENWELL.

His Majesty's Privy Council having approved of precautions proposed by the Board of Health in London, on the alarming approach

OF THE

INDIAN CHOLERA

It is deemed proper to call the attention of the Inhabitants to some of the Symptoms and Remedies mentioned by them as printed, and now in circulation.

Symptoms of the Disorder;

Giddiness, sickness, nervous agitation, slow pulse, cramp beginning at the fingers and toes and rapidly approaching the trunk, change of colour to a leaden blue, purple, black or brown; the skin dreadfully cold, and often damp, the tongue moist and loaded but flabby and chilly, the voice much affected, and respiration quick and irregular.

REMEDIES;

All means tending to restore circulation and to maintain the warmth of the body should be had recourse to without the least delay.

The patient should be immediately put to bed, wrapped up in hot blankets, and warmth should be sustained by other external applications, such as repeated frictions with flannels and camphorated spirits, poultices of mustard and linseed (equal parts) to the stomach, particularly where pain and vomiting exist, and similar poultices to the feet and legs to restore their warmth. The returning heat of the body may be promoted by bags containing hot salt or bran applied to different parts, and for the same purpose of restoring and sustaining the circulation white wine wey with spice, hot brandy and water, or salvolatile in a dose of a tea spoon full in hot water, frequently repeated; or from 5 to 20 drops of some of the essential oils, as peppermint, cloves or cajeput, in a wine glass of water may be administered with the same view. Where the stomach will bear it, warm broth with spice may be employed. In every severe case or where medical aid is difficult to be obtained, from 20 to 40 drops of laudanum may be given in any of the warm drinks previously recommended.

These simple means are proposed as resources in the incipient stages of the Disease, until Medical aid can be had.

THOS. KEY,
GEO. TINDALL, *Churchwardens.*

Sir Gilbert Blane, Bart. in a pamphlet written by him on the subject of this Disease, recommends persons to guard against its approach by moderate and temperate living, and to have in readiness, the prescribed remedies; and in case of attack to resort thereto *immediately* but the great preventative he states, is found to consist in a *due regard to Cleanliness and Ventilation.*

N.B It is particularly requested that this Paper may be preserved, and that the Inmates generally, in the House where it is left may be made acquainted with its contents.

NOV. 1st, 1831.

T. GOODE, PRINTER, CROSS STREET, WILDERNESS ROW

CHOLERA-PREVENTION POSTER, LONDON, 1831
Typographic design became full of possibilities, as shown on this poster: it was circulated in London to inform people about the impending epidemic, its symptoms and remedies.
Wellcome Institute Library, London

Printing with stone and light

The story goes that, in Bavaria, the playwright Aloys Senefelder (1771–1834) was about to leave the house one day: his mother dictated a shopping list, which he happened to write on stone with a grease pencil. He had been searching for a cheap way to copy his work, and observed that the marks on the stone retained the inks, by affinity with their fatty component; he then developed this method for printing.

Senefelder made a series of coloured lithographs (*lithos* means 'stone' in Greek), and predicted that some day his method would be used to reproduce paintings. The French Baron Lejeune, who was stationed nearby, brought the method back to France and suggested using it for promoting Napoleon's campaigns.

LITHOGRAPHIC SELF-PORTRAITS, RENÉ LAENNEC, NINETEENTH CENTURY

The surface and texture of a lithographic print can closely approximate that of paint itself: the graininess comes through almost as if it were real skin. These self-portraits were made by René Laennec, the French physician who invented the stethoscope and eventually died from tuberculosis.

Bibliothèque Nationale de France. © Roger-Viollet (left)

The importance of lithography was immediately evident to Francisco de Goya (1746–1828), Eugene Delacroix (1798–1863), and Théodore Géricault (1791–1824). As the artist draws on a smooth porous surface, velvety textures and colour variation produce a painterly effect.

Artists took up lithography as an ideal means of experimenting with flattening the picture plane, destroying the illusion of spacial recession. The technique, although cumbersome – each colour requires its own stone-impression – is cost-effective since the number of prints that can be produced is unlimited. Stone endures even after copper wears out.

Thus, for economic reasons, lithography also appealed to political cartoonists, who soon employed it to good effect in newspapers and magazines.

Lithography was widely used, from illustrations in popular books to trademarks and political campaigns. Senefelder had correctly guessed that his invention would play a crucial role in the communication arsenal, and has done to this day.

'Nadar Elevating Photography to the Height of Art', NINETEENTH CENTURY

Nadar, who gave up the study of medicine for photography, became the first aerial photographer, and pioneered photography using artificial light. As cartoonist and lithographer, Nadar had an ambition to portray all the famous figures of his time. Only the first page of his *Pantheon Nadar* was published; it was a commercial failure. Still, it leaves us with memorable caricatures of over two hundred celebrities.

Courtesy George Eastman House

Visually effective technology is a highly competitive arena in which a new goal was soon set: access for all to reproduce images by means of photography. The function of a dark chamber (or box), with an aperture, had been known since the time of Aristotle. When light passes through the aperture, it casts an inverted picture on the opposite wall. Like many discoveries, it took centuries to turn it into a practical application.

The small portable chamber, or camera obscura, had been commonly used by artists since the Renaissance, and its opening later incorporated a lens. But no method, other than printing, had been found for 'fixing' the image.

The missing link, the fact that silver nitrate darkens on exposure to light, had been described as early as 1614. Thomas Wedgwood (1771–1805), the son of the famous ceramist, Josiah, while experimenting with his father's camera obscura, put leather sheets treated with silver nitrate inside the box. By doing so, he developed the technique of chemical fixing of images. When the light struck the sheets, it imprinted them with the image seen through the aperture.

Various media were subject to intense experimentation. Joseph Nicéphore Niepce (1765–1833), a research chemist and lithographic printer, invented heliogravure, using sunlight (*helios* means 'sun' in Greek), by a process for reproducing images which required a full day's exposure.

Louis Jacques Daguerre (1789–1851), a French actor and skilled painter who contributed to the invention of the diorama – huge circular paintings illuminated in dark rooms that attracted the whole of Paris – teamed up with Niepce to develop the daguerrotype. The latter, a silver image printed on a copper plate, became the basis of modern photography. Even though daguerrotypes were fragile, and yielded only a single print, within the year after the invention's public debut, half a million daguerrotypes were produced.

Meanwhile, in England, William Henry Talbot (1800–77), a distinguished archaeologist and mathematician who enjoyed sketching landscapes but was frustrated by his inability to capture the changing nature of light, saw photography as a solution. He devised practical photographic processes, plates and paper: his negative, from which an infinite number of prints could be made, reworked and enlarged, was revolutionary.

Exposure time was gradually reduced and fixing – initially involving wet plates that required a considerable time to dry –

JOHN HERSCHEL, ALBUMIN PRINT BY JULIA MARGARET CAMERON, 1867

Herschel played a crucial role in the development of spectroscopy and coined the word 'photograph' by analogy to the telegraph, already in use. Cameron, befriended by both artists and scientists, having received photographic equipment as a 49th birthday present, made penetrating portraits.

Courtesy George Eastman House

was also improved upon. Switching from wet plates to dry plates liberated photographers just as oil paint in tubes had released artists from their studios.

Most artists avoided photography, at least officially, identifying it as a threat to their mission: to represent the world around them. It did have its art-world champions, however. In particular, the English critic John Ruskin (1819–1900) saw the creative potential: 'Daguerrotype helps the artist accomplish the reconciliation of true aerial perspective and chiaroscuro with the splendour and dignity of elaborate detail.'

In France, scientists helped to raise the money to support research in photography, and promoted it at the Académie des Sciences as well as at the Académie des Beaux Arts – artists rarely got involved. Yet, Ingres (1780–1867) the French master of representation, admitted: 'Which one amongst us is capable of such truthfulness, of such firmness when it comes to interpreting delicate lines and shapes. Photography is beautiful, but we should not tell anyone.'

In the United States, the visionary Alfred Stieglitz (1864–1946), established photography as an art in its own right. The European painting avant-garde, with Duchamp,

Moss photographic department, 1877
Reporters on the rise of photoengraving revealed that such reproductions had been published together with hand engravings for ten years. Readers had not noticed the difference.
Scientific American

Matisse and Picasso, made a spectacular debut on the New York scene thanks to Stieglitz. An artist trained as an engineer, he felt comfortable with mechanical tools used in art: 'If only people would broaden their concepts to include concern about the brotherhood of man and the machine, the world would be a great deal better.'

Photography developed new ways of documenting reality and invited people to see things that they had never before imagined. The result was a remarkable expansion of visual horizons.

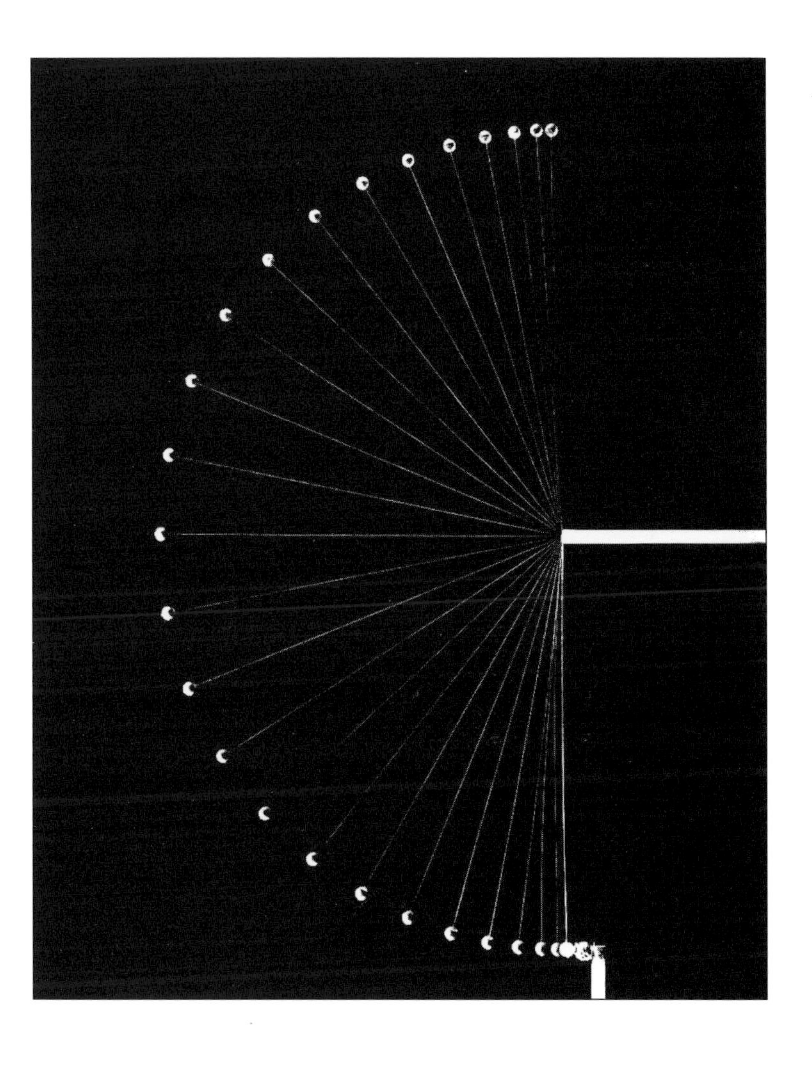

Transformation of Energy, Berenice Abbott, Cambridge, Massachusetts, c. 1958
Abbott's photographs were aimed at illustrating the laws of physics. Her works were scientific in their intent and aesthetic in their effect.
Commerce Graphics Ltd, Inc.

Communicating through design

In reaction to the machine age, the Arts and Crafts movement in England, calling for a revival of nature and individual expression, had an extraordinary impact on graphic design: it was followed by Art Nouveau on the continent.

The latter decorative style, representing plant and animal motifs stylized in the extreme, was, to some extent, an artistic commentary on Darwinism (Darwin as it happens also published work on plant morphogenesis). The style later developed a geometric phase reflecting the changing relationship with space.

A new approach advocated that all branches of art, from graphics to painting, and from sculpture to design, shared values expressed by mass-produced quality goods. The German

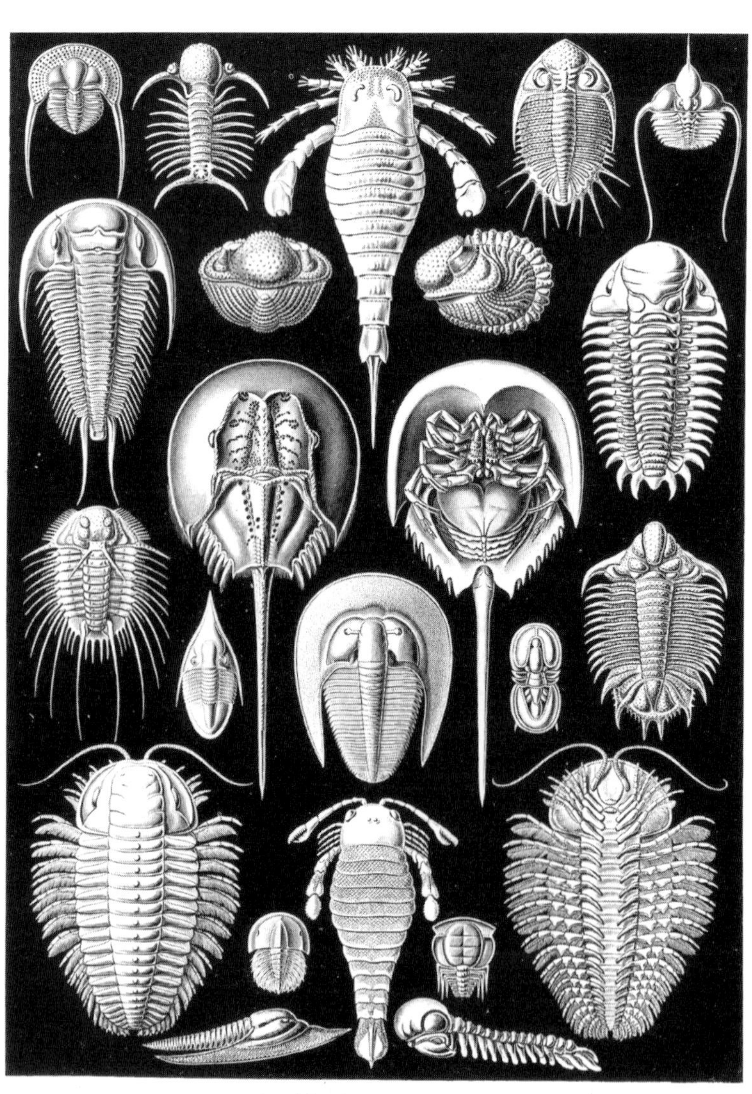

Aspidonia. — Schildtiere.

TRILOBITES, EURYPTERIDS AND HORSESHOE CRABS, ERNST HAECKEL, C. 1904

Haeckel was the first German biologist to become an ardent Darwin partisan and the first to use the word 'ecology'. The connection between his plant and animal drawings and the rise of Art Nouveau is evident. His work fascinated artists like Gallé, Guimard and Mucha.

Artforms in Nature. *Württembergische Landesbibliothek. Photo Joachim Siener*

Bauhaus school, for example, invested its creative energy into 'total design', implicitly recognizing the common roots of fine and applied arts.

Cubists, Futurists and Constructivists worked in parallel with philosophers and poets. The need to harmonize all facets of life was stressed by Aldous Huxley (1894–1963): 'It is obvious that the machine is here to stay. Whole armies of Morrises and Tolstoys could not now expel it. Let us exploit them to create beauty – a modern beauty while we are about it.'

Machines literally flooded the world with information: communicating masses of messages to culturally diverse audiences through simple symbols became the norm. In this context, a kind of graphic Esperanto was designed by Otto Neurath (1882–1945), a Viennese philosopher with a passion for hieroglyphs. His modern pictographs became the prototypes for international signage: the 'world language without words' spread out around the globe.

Design is now applied to everything, from typefaces to fashion and home furnishings. International events of any kind require an identity, or a logo. Design is obviously not a modern invention but the mass media have turned it into a global phenomenon.

DEMOCRACY LOGO, JERZY JANISZEWSKY, POLAND, 1980

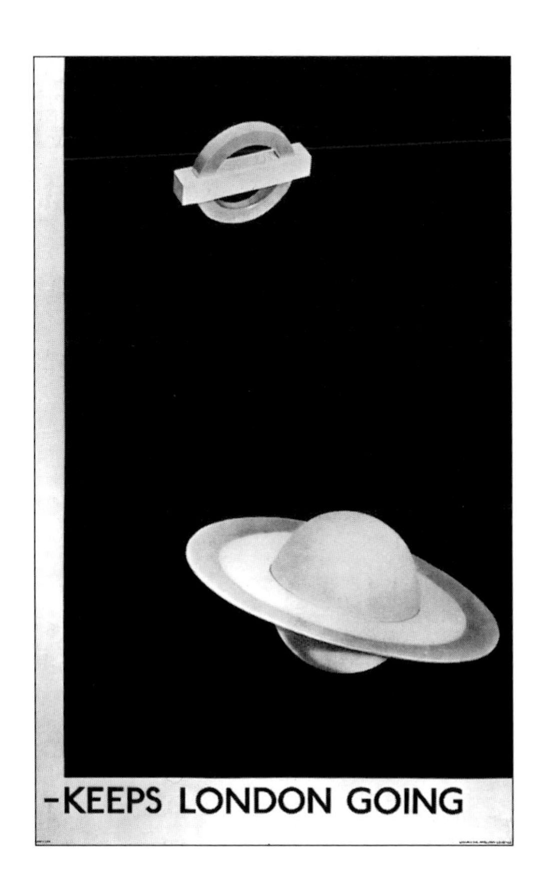

KEEPS LONDON GOING, MAN RAY, 1932

Man Ray, who 'painted with light', enlarged the photographic media range by creating solarizations and rayographs. He discovered the latter technique by chance, having left small objects such as keys and cotton on photographic paper which produced mysteriously poetic images. This poster designed for the London underground is based on an analogy with Saturn.

Offset lithograph printed in colour, 39 5/8" × 24 1/4". # K 9149.
The Museum of Modern Art. Gift of Bernard Davis.
Photo © 1998 The Museum of Modern Art, New York

Creation of virtual worlds

While photocopying machines greatly accelerated the spread of information and images, and even generated 'copyart', computers and networks effectively engaged the most disparate cultures in dialogue.

With satellites, works of art in remote museums and scientific data in distant laboratories are instantly accessible. The interactive technologies, synthesizing sound and pictures, are burgeoning fields of endeavour where the public is often informed before the specialists!

It can hardly be a coincidence that Douglas Hofstadter, an expert in artificial intelligence, had already revealed in the 1960s a profound interest in aesthetics. His Pulitzer prize-winning book, *Gödel, Escher, Bach: An Eternal Golden Braid,* devoted to correspondences between mathematics, graphic art and music, demonstrates how systems can be expressed either as theorems, drawings or musical notes.

Even though artificial intelligence has greatly advanced, it is still in its infancy; where creativity is concerned, computer intelligence cannot compete with the complex functioning of a young child's brain. It is difficult to guess what is coming next, however. Computers already recognize voices and may well make the typing keyboard obsolete . . . will writing still be needed?

ORIGINAL MACINTOSH ICONS (1984)

MacDraw MacWrite MacPaint

ICONS CREATED FOR THE 128K MACINTOSH COMPUTER, SUSAN KARE AND BILL ADKINS, 1984

Design of bankcards or computer icons requires the collaboration of a wide variety of experts: economists and sociologists as well as artists and technologists.

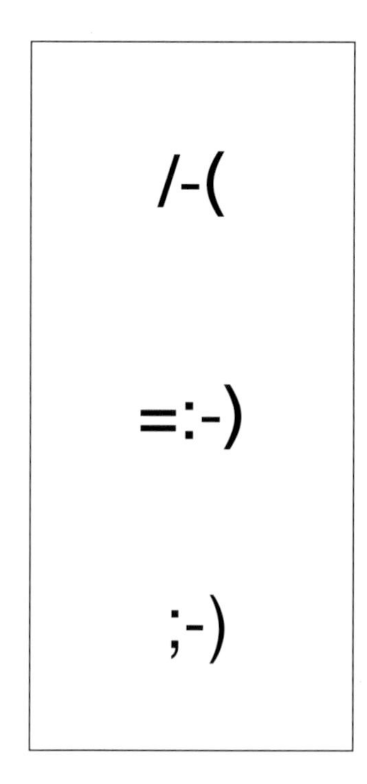

'EMOTICON', LANGUAGE USED BY CYBERSURFERS

The language used on the Internet resembles ancient pictograms. Reading from top to bottom: rather than being angry, I express my joy . . . and send you a 'wink'.

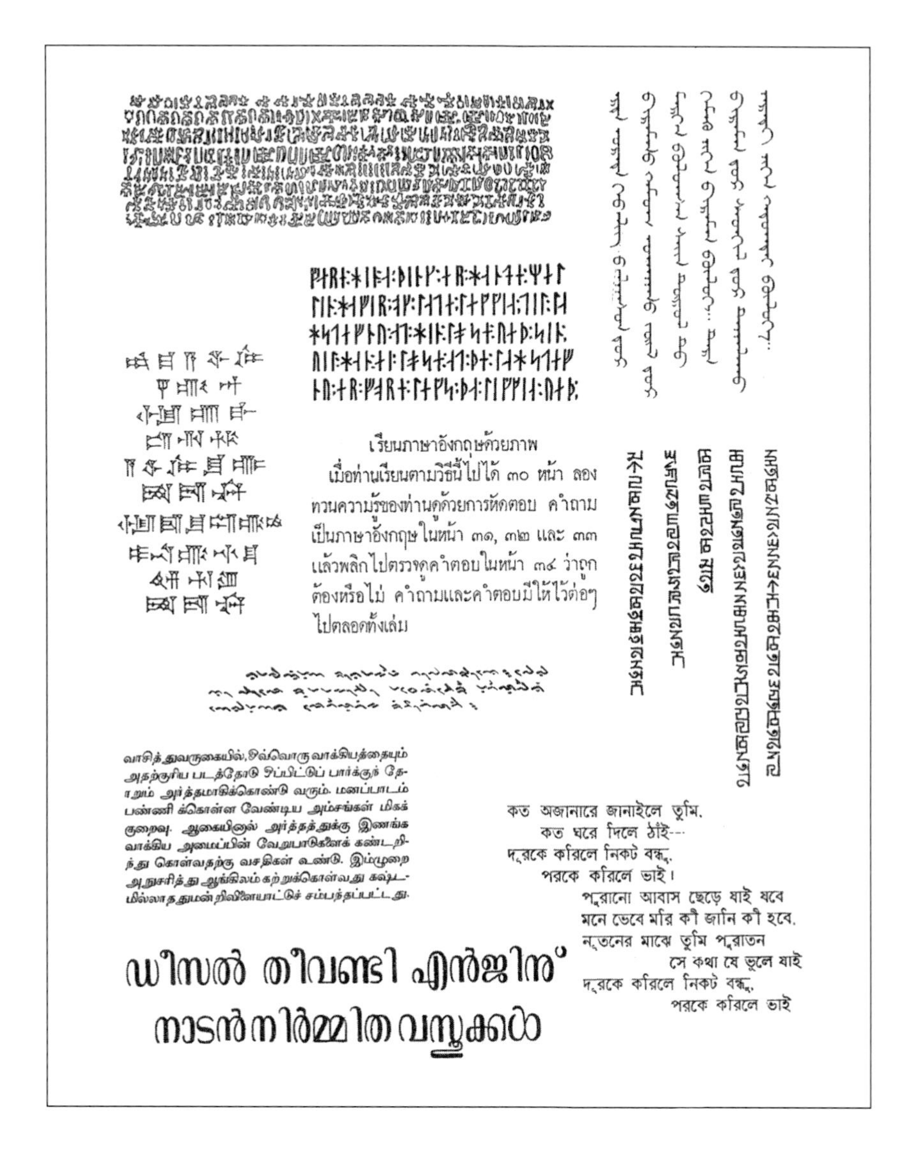

COLLAGE OF SCRIPTS, DOUGLAS HOFSTADTER

Hofstadter said of the above languages: ‘As an outsider, I feel a deep sense of mystery as I wonder how meaning is cloaked in the strange curves and angles of each of these beautiful aperiodic crystals. In form there is content.’

Gödel, Escher, Bach, *1979. Courtesy of HarperCollins Publishers*

MATISSE DRAWING A *DANCE* PANEL, 1931
One version of the *Dance* was created for a specific setting in the home of Albert Barnes, American collector and arts patron. After almost a full year's work, Matisse had to begin again because he had been using the wrong measurements. Computers would have greatly expedited the recalibrating process. The art of Matisse, the 'omission genius', for whom each line or space occupies a highly specific place, has a metric character.

Surfing on the Web and playing with DVDs – (digital versatile disks) which can hold the content of several CD-ROMs (compact disk read-only memory) – structure thought in such a way that one can travel freely within a huge body of knowledge. This perhaps signals a move away from the linear thinking painstakingly developed over the past five hundred years.

We are about to enter a new millennium with as many unanswered questions as our nineteenth-century forebears raised when Darwin formulated his theory of evolution; painters bade farewell to conventional representation and beauty as the ultimate aims of art. What is more, in the eighteenth century, machines were already expected to replace humanity . . . yet we are still here.

Since the dawn of civilization, artists and scientists have mutually contributed to the cause of visual communication. From abstract, pictorially imaginative writing systems to illuminated manuscripts, displaced by printing, and onwards to photography and computer technology, art and science have been indissociable.

LILY'S ADVENTURES IN COMPUTER LAND,
LILLIAN SCHWARTZ, 1998
The author of this computer-created film said: 'I created a Matisse to break the background into larger areas and placed Modigliani's *Nude* amid Matisse's *Dance*. I abstracted the background more, the figures became silhouetted and the *Nude* was cut out of the background.' One could speculate that computers will assist in calculating, posthumously, paths that creators might have taken.

Technique and the performing arts

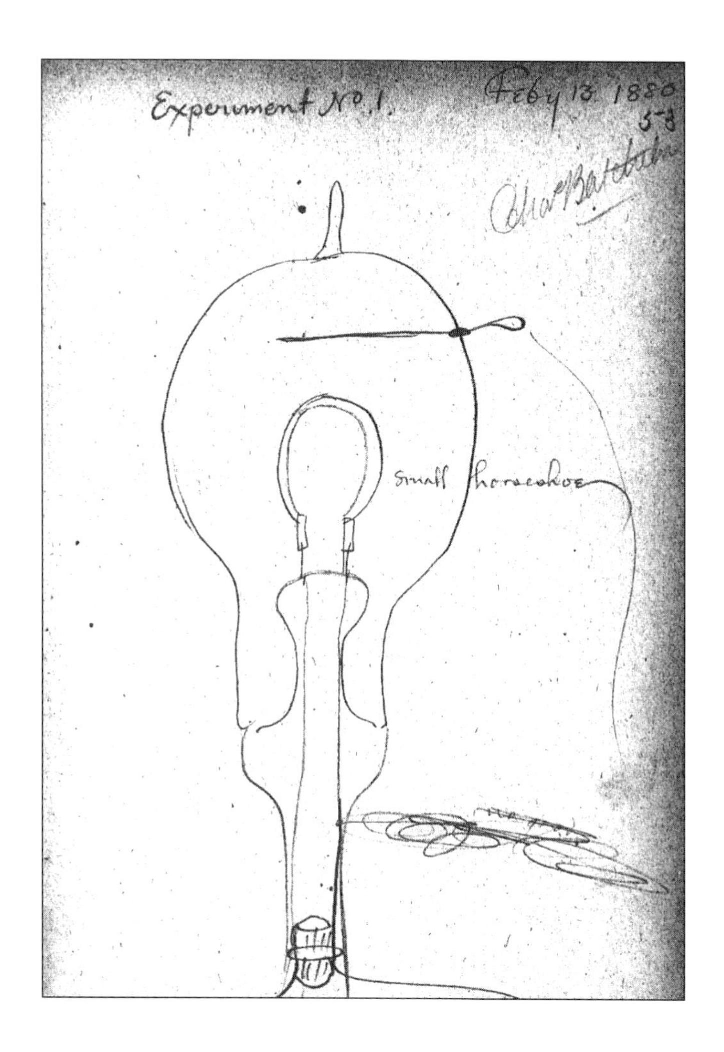

LIGHT BULB DRAWN BY THOMAS EDISON AT HIS RESEARCH LABORATORY, MENLO PARK, NEW JERSEY, 1880

Edison widened audiences for the performing arts by introducing the phonograph and the cameraphone – which synchronized the phonograph with a film projector – and by laying the foundations for the development of transistors, later used in radios. In order to make a bulb that provided light for prolonged periods of time, hundreds of types of filament were tested.

US Department of the Interior. National Park Service. Edison National Historic Site

Waves of light and sound

As men started to build, ritual forms of expression brought architectural spaces to life. Light is the orchestrating element of the performing arts, as it is that of architecture. Artists have long endulged in combining waves of light, and waves of sound, with other forms of higher order. Goethe, for example, described architecture as 'frozen music'.

Whether light is made up of particles or other substances was already being debated at an intuitive level in Ancient Greece, and philosophers throughout history have been fascinated by the nature of light. The discussion was revived by Leonardo who stated that it was constituted of waves.

It was discovered in the nineteenth century that light is made up of electromagnetic waves. Then, Heinrich Herz

Fire, Giuseppe Arcimboldo, 1566
Arcimboldo, a Milanese Renaissance painter employed by the Emperor Rudolf in Prague, invented a form of musical notation using colour and tone, based on Ancient Greek theory. The painters Kandinsky and Mondrian, and the composers Telemann and Cage, studied the relations between colour and music.
Kunsthistorisches Museum, Vienna

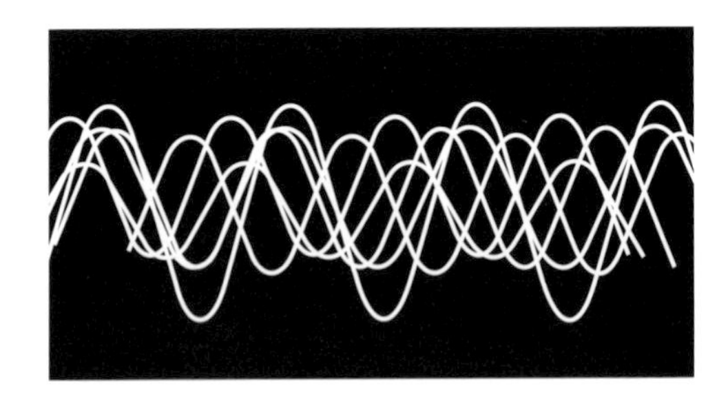

White light

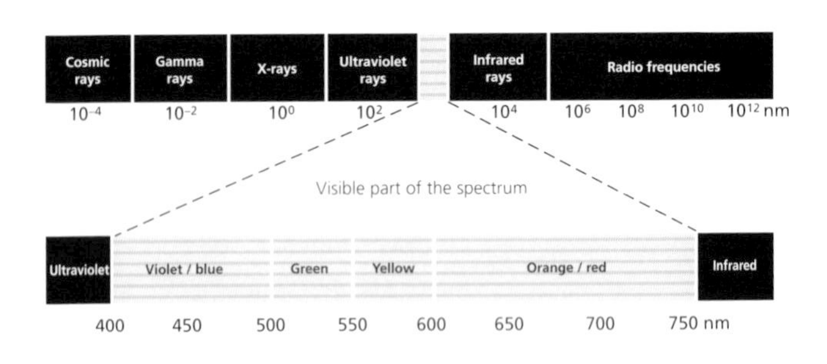

Diagram showing various types of wavelength
White light is made up of wavelengths of different colours; visible colour is only a small part of the spectrum. Laser light consists of identical concentrated waves and thus produces an extremely bright light.

Laser waves

(1857–94) detected a non-visible wave that had a much longer wavelength than visible light and called it a 'radio' wave. He could not have foreseen, however, its importance in the broadcasting revolution that was soon to occur.

The transition really began when Guglielmo Marconi converted sound into a current that travels silently, transmitted 'wirelessly' when sent into space (1894). Antennas receive and reconvert them into sound waves; broadcasting became instantly popular.

The almost simultaneous invention of the moving picture resulted later in combining waves of light with waves of sound. Physicists today accept the paradox that light can take many forms, behaving sometimes as particles, sometimes as waves. Meanwhile, artists have turned waves into a symbol of the performing arts.

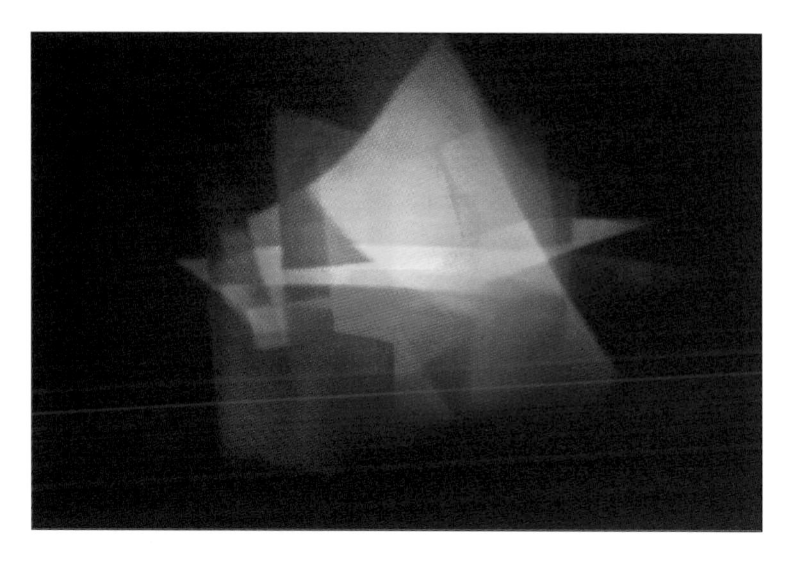

LIGHT MILL, COMPUTER-GENERATED HOLOGRAM, DIETER JUNG, 1987
Laser light is split into a reference beam and an object beam; the two beams form an interference pattern whose record is called a hologram. Holography is a medium that establishes a dynamic relationship between the artist, the artwork and the public.
Photo Dieter Jung

Vibrations of the soul

Many aspects of the history of music parallel the development of the visual arts. Some cave paintings already suggest links between visual composition and cave acoustics. Since the dawn of humanity, people have produced sound waves from shells, animal skins, bones and horns, wood and stone.

We have explored materials for sound, perhaps to satisfy spiritual needs. Our visceral love of music may originate in the biological rhythms of the heart or of our breathing system. (Medieval music notational signs were called [p]neuma.)

The Greeks, faithful to their rational approach, found that harmony of numbers had a special appeal. The mathematician and physician Pythagoras (560–480 B.C.) who believed in the therapeutic virtue of music, found that the tone produced by plucking a string changed in proportion to its dimensions.

Dividing the strings by whole numbers, Pythagoras could produce half the notes of an octave, indicating that a sequential relationship was at the basis of musical harmonies – the seven notes, like the seven planets, were connected to mystical forces. Despite the sophistication of their theories, Greek music probably remained monodic, with all notes sung in unison.

MECHANICAL THEATRE, HERO OF ALEXANDRIA, FIRST CENTURY A.D.

The Ancient Greeks made automates and organs. Hero of Alexandria is credited with such inventions, including a steam engine. Soon after Constantine V presented Pepin Le Bref (757) with a pipe organ – a symbol of power because of its advanced technology – the instrument was adopted by the Catholic churches of Western Europe.

Photo Bibliothèque Nationale de France

Gregorian chant, or plainsong – named after Pope Gregory the Great – developed to unify the sacred music of the Western Church, was an amalgam of Judeo-Christian psalms and antique hymns. In the beginning, different singers sang the same music at similar levels within a narrow range. There was no fixed rhythm and music followed the words of the prayer. Progressively, multi-voice singing at different levels, or polyphony, appeared. It seems likely, however, that polyphony existed long ago in Africa.

In the West, the Carolingian period saw the development of musical notation written on a single staff form. The path of each voice could be charted and several voices could sing different musical lines simultaneously. The music notation system, *doh-ray-mi-fah-soh-lah-te,* was invented by Guido d'Arezzo (995–1050); he attributed the first syllables of a phrase in a popular hymn to the ascending music tones, as a memory aid.

Four- and five-line musical notation appeared, allowing for sophisticated forms of polyphony. The Church resented the proliferation of instruments and of polyphony. Yet a brilliant synthesis of ecclesiastic and secular music was achieved by the Italian 'Prince of Music', Giovanni Perluigi da Palestrina (1525–94).

Audacious composers formed secret organizations such as the Camerata, so-called because they met behind closed doors (*camera* means 'vault' in Latin), of which Galileo's father was an active member, being a musician and experimenter in acoustics. Mathematicians and astronomers such as Kepler, Galileo and Descartes were versed in music.

As gifted singers developed their singular talent and their autonomy, the demand for solo-voice music grew, culminating in a new musical form, the opera. Claudio Monteverdi (1567–1643), nicknamed Il Divino, was the first composer to treat it as a major part of his endeavour. Within a few decades dozens of opera houses sprang up in Italy.

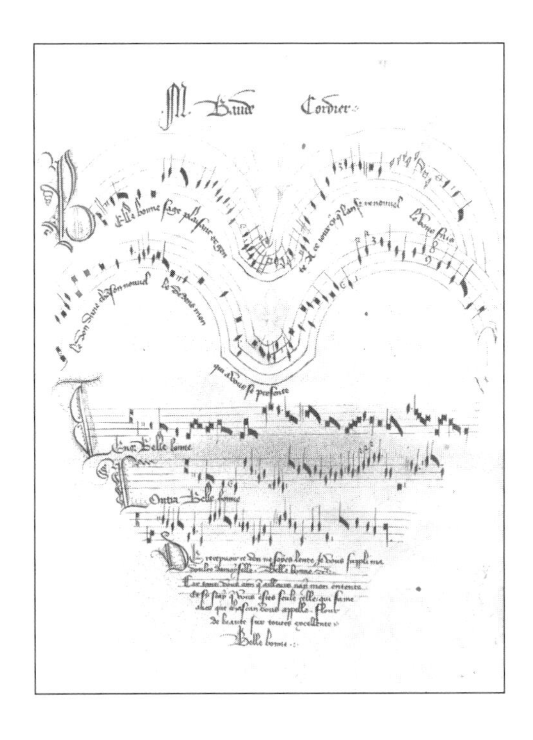

'HEART,' IN BALLADS, MOTETS AND SONGS,
BAUDE CORDIER, EARLY FIFTEENTH CENTURY
'Beautiful, wise, kind and pleasant', this elaborate piece of music was probably given with the heart to a lady.
Musée Condé, Chantilly. Giraudon

SYSTEM OF MUSICAL NOTATION,
JOHANNES KEPLER, 1619
Like the Ancient Greeks, Kepler believed in a relation between the planetary order and music. In his treatise *The Harmonies of the World,* musical scales correspond to the sounds thought to be produced by the planets' movements. Kepler wondered 'which planet was singing the alto, which one the soprano, the tenor or the bass'.
QB41.K42. Library of Congress, Washington, D.C.

The orchestra developed around the same time, and Monteverdi, who had a brilliant reputation in every type of music, conducted the first operatic performance, *Orfeo* (1607), in which the instruments – of which there were forty! – and their individual roles were specified.

Instrumental music had long been restricted to the popular domain, with the notable exception of the pipe organ. With the emergence of scientific instruments, religious music was united with musical instrumentation. The Renaissance saw a convergence of technique and science, and in particular, of music and theory. One-point perspective in painting and architecture had its musical counterpart in the key introduced to unify composition.

Polyphony was further pursued with the rigours of science. Meter, harmony and counterpoint – the interplay of independent musical strands – were developed 'quasi-mathematically'. In his *Traité d'Harmonie Universelle,* the French theologian and mathematician, Marin Mersenne (1588–1648), stated that 'the science of music depends on arithmetic and geometry, and also on physics, from which it borrows knowledge of sound, and its causes which are the movements, the air, and the bodies that produce sound'.

Mersenne, 'Europe's mailbox', exchanged knowledge with scientist-philosophers such as René Descartes (1596–1650) and Blaise Pascal (1623–62), in France, and the Italian or Spanish Jesuit missionaries travelling in the the Far East. Mersenne gave a full description of contemporary instruments, while new ones rapidly appeared.

The Arabs brought kettle drums and trumpets, luths and guitars to the West. The contributions from instrument-makers to the evolution of music can never be overstated: numerous compositions stress virtuosity and aim to show both performer and instrument to their best advantage.

The Vivaldi and Bach dynasties considered themselves to be above all dedicated technicians. During his lifetime, Johann Sebastian Bach (1685–1750), was better known as an organist than as a composer. Until the nineteenth century, music was created on request, often for a single performance.

The trend for new instruments was heightened by the Eastern influx into Venice: Greek emigrants kept up with the antique interest in automates. Around the fifteenth century, the harpsichord, a complex keyboard instrument that produced

CHINESE BELL, SIXTH CENTURY B.C.

In China, bells were used as precision instruments for weighing valuable commercial goods. The tones emitted by bells are directly related to the size, volume and thickness of their walls; producing a bell of a certain tone requires technical knowledge in metallurgy. The Chinese also formulated advanced musical theory. As early as the sixteenth century, they knew about the 'equal temperament' – a scale of notes with equal intervals, which was developed only recently in the West.

DETAIL OF THE MUSIC CHAMBER CEILING, NORFOLK HOUSE, UNITED KINGDOM, EIGHTEENTH CENTURY
Boethius (480–524) invented the Quadrivium, an educational programme that covered the four mathematical disciplines of arithmetic, geometry, anatomy and music, which was influential well into the Renaissance.

sound by the mechanical plucking of strings, was introduced. It was joined by local variants, such as the virginal and the spinet, and improved by the addition of half-tones and several octaves. But there was a serious limitation: the player could not modify the intensity of the sound of an individual string, since they were all plucked with equal force.

New sounds produced by new instruments engendered new musical forms: the sonata, or 'sound piece', was the term used to designate various types of music for specific instruments. Concertos and chamber music became fashionable; three- to four-movement symphonies appeared, and overtures were introduced at the beginning of operas.

The demand for excellence increased as musicians started to travel from court to court, and printed musical scores began to circulate. Little, if any, of Bach's music was published while he was alive. (This explains, to some extent, why Joseph Haydn (1732–1809) wrote over a hundred symphonies, Wolfgang Amadeus Mozart (1756–91) half as many, and Ludwig van Beethoven (1770–1827) only nine.)

In the realm of mechanical progress, an innovative string instrument, the pianoforte (1709), allowed for tonal variation. In this ancestor of the piano, strings were hammered instead of

GLASS HARMONICA INVENTED BY
BENJAMIN FRANKLIN, GERMANY, EIGHTEENTH CENTURY

Franklin called music 'that charming science'. Having heard the *vérillon* – a set of musical glasses filled with graduated amounts of water to produce various notes when their rims were traced by moistened fingers – he designed a mechanical version that produced thirty-seven tones. The fashion lasted for a century. Mozart and Beethoven both composed pieces for such instruments.

Musée des Instruments de Musique de Bruxelles.

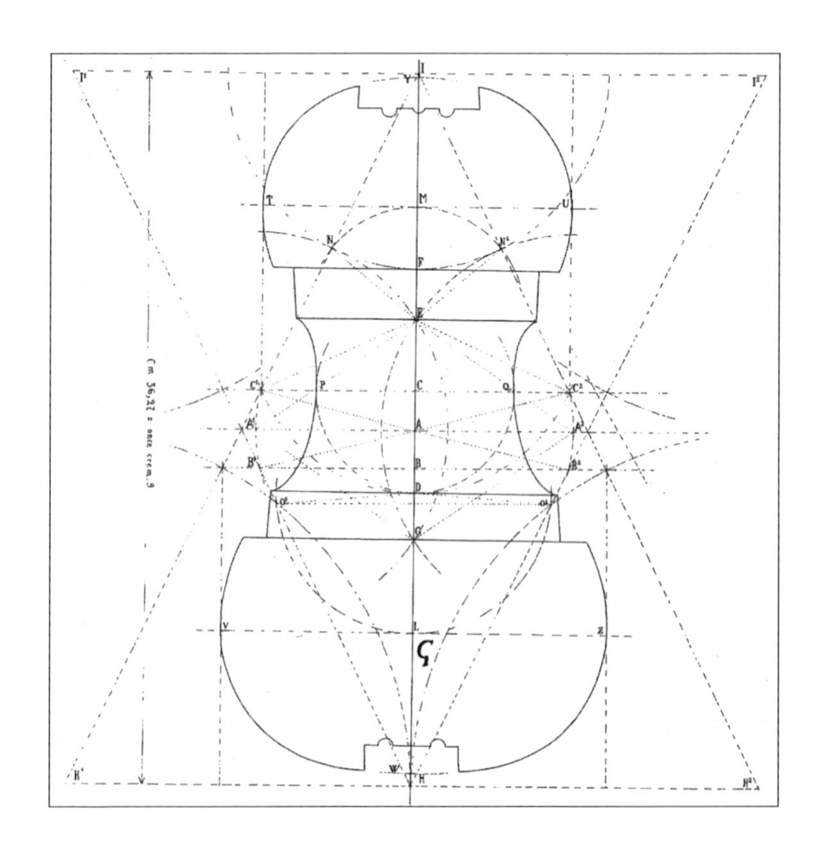

ANALYSIS OF THE PROPORTIONS OF A STRADIVARIUS,
SIMONE SACCONI, 1972

Violins have aesthetically appealing shapes, yet their proportions are rigorously defined. Ancient instruments were built from as many as a hundred pieces of special wood treated and assembled in time-honoured ways. The timbre of rare violins can now be reproduced on carbon-fibre instruments, whose shape and sonority are tested with advanced computer technology. Perhaps affordable violins will soon be manufactured in carbon-fibre – a material used in hi-tech sports and aeronautics equipment.

The Secrets of Stradivari. *Libraria del Convegno*

plucked, with a far greater range of expression than its predecessors. The piano substituted a scherzo for a minuet.

Mozart owned a pianoforte and was so interested in it that he had a special pedal constructed for his improvizations. The harpsichord and the piano would compete for over a century. Those who loved the plucked string found the piano's sound rather brutal, until compositions of Frédéric Chopin (1810–49) and Franz Liszt (1811–86) conquered the public ear. The iron frames of pianos were then produced by the latest technologies. The volume of sound increased substantially and turned the piano into a truly popular instrument.

Around the same time, Hermann von Helmholtz (1821–94), physicist, physiologist and pianist, published *The Doctrine of the Sensation of Tones.* He discovered that he could make piano strings vibrate by vocalizing or using a magnet, and produced corresponding vibrations in a tuning fork. By comparing sounds produced in different systems, he developed a scale and a theory for explaining the physiology of hearing. This time science had encroached upon the realm of music.

The orchestra was eventually divided into sections – strings, woodwind and brass – like an experimental chamber where sound was tested on a large scale. Acoustical engineering made its architectural debut in the design of Boston's Symphony Hall (1868).

At about this time, Richard Wagner (1813–83), also a talented painter and poet, while attempting to reconcile antique mythology with modern music, commissioned specially adapted instruments. The development of new materials, such as brass alloys, fully benefited Wagner's chromatically dissonant music – resembling Cézanne's compositions. Meanwhile, Claude Debussy (1862–1918), content with traditional instruments, experimented with an Impressionist style.

At the turn of the twentieth century, new concepts in music, as well as in science, took the lead over technology. The tenets of classical physics, based on two-dimensional geometry, were transposed into multiple dimensions. This was accompanied by a dramatic break in the history of music. The Austrian-born composer Arnold Schoenberg (1874–1951) introduced atonal compositions – in which each note had the same relative importance as all others – written within a twelve-tone scale, for which Cubism was the visual equivalent. During his youth, Schoenberg baffled his entourage by a series of hallucinatingly

Tympanum Player, Kintzing, eighteenth century
This instrument has forty-six chords and plays eight different tunes. Musical automates were fashionable at a time when there was little distinction drawn between art, craft and technical achievement.

Sound analyser, Koenig, nineteenth century
Helmholtz, one of the greatest scientists, contributed to improving the acoustics of the Steinway piano. By using the sound analyser, he deduced the mechanism of auditive physiology and conceived an otoscope (a medical device to look inside the ear.) His work on the physiology of sight resulted not only in theory but also in the ophthalmoscope (an instrument for examining the eye), still used by doctors.

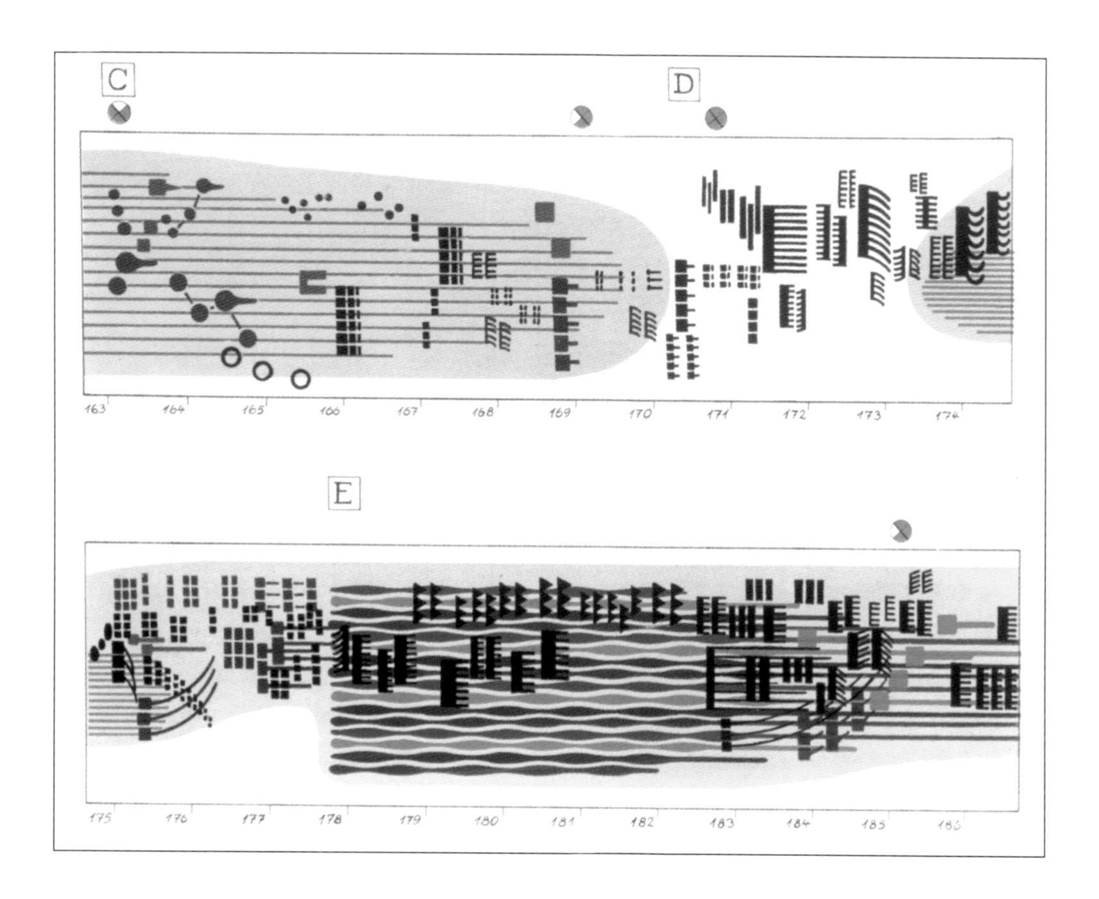

ARTICULATION, GYÖRGI LIGETY, 1958
Today, Westerners painstakingly write complex music similar to that composed in Africa since time immemorial.
Courtesy of Schott Muzik International, Mainz. Photo Bibliothèque Nationale de France

expressive self-portraits. Yet he suddenly abandoned painting to devote his entire research to music.

Igor Stravinsky (1882–1971), another of the most innovative twentieth-century composers, created revolutionary music that, to some, still sounds 'extra-terrestrial'. But even Stravinsky portrayed himself as just 'a craftsman whose materials of pitch and rhythm in themselves harbour no more expression than a carpenter's log'; his work-table resembled that of a surgeon and the precision of his scores resembled that of a mathematician.

The spiritual heirs of Bach, Beethoven, Schoenberg or Stravinsky composed, embellished and revisited existing themes within a tradition – an approach also used by the most creative painters.

In an era in which radar, capturing cosmic sounds from the far reaches of the universe, can be exploited for musical purposes, there seems to be no limit to the roles of science and technology in the development of music. In many works, conventional instruments have been replaced by digitally determined sounds: visual art tools have gone the same way.

Iannis Xenakis, the composer, mathematician-engineer and visual artist who collaborated with Le Corbusier, said: 'In music, numbers and mathematics are the founding elements of its universality.' This quote might have come from one of Pythagoras' dreams.

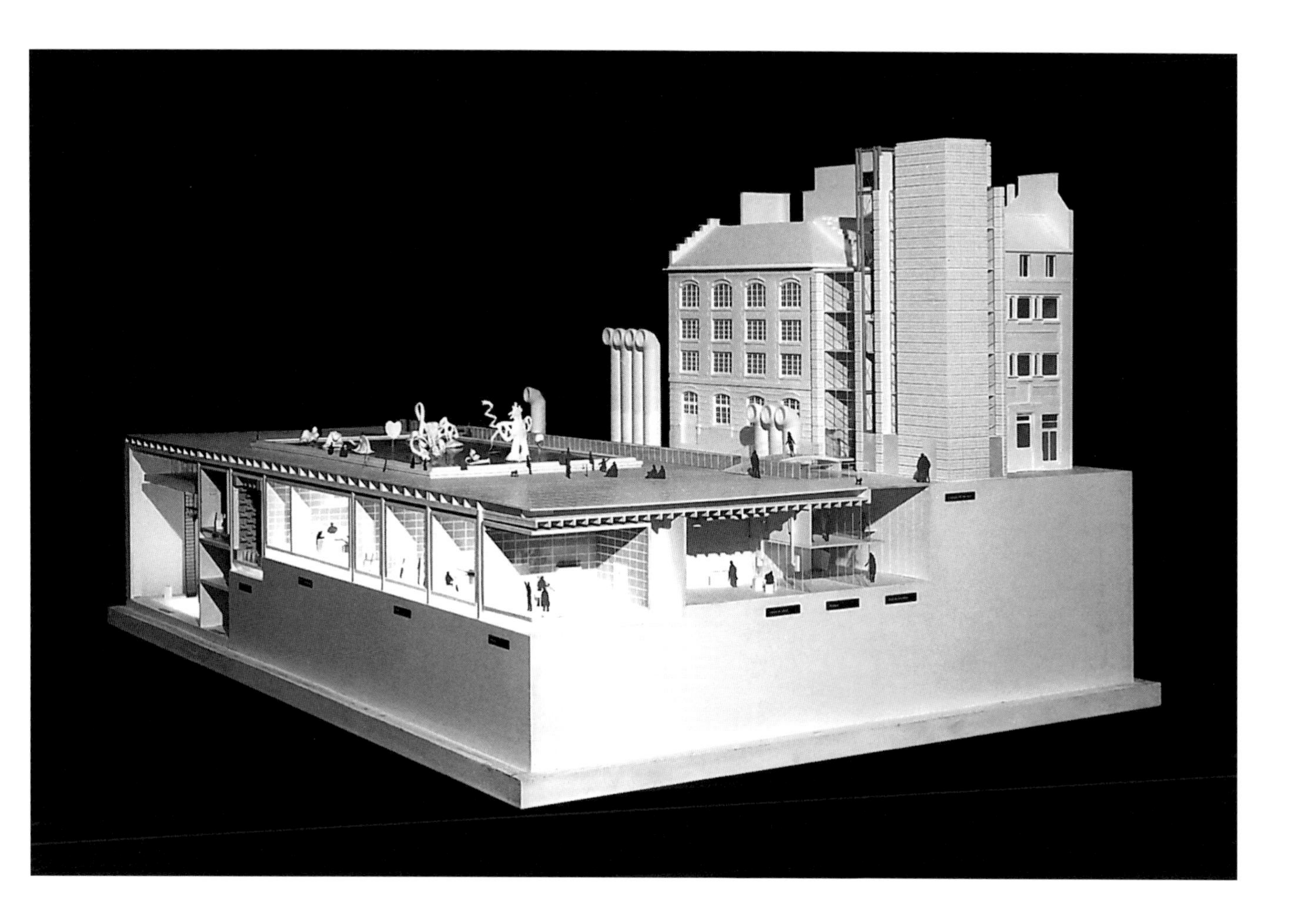

MODEL OF IRCAM

Adjacent to the Pompidou art centre in Paris, underneath mobile sculptures, is the facility known as IRCAM (Institut de Recherche et de Coordination Acoustique et Musicale). Founded in 1977 under the direction of the composer/conductor, Pierre Boulez, it became a laboratory for new techniques and compositional forms. Concerning music made of permutable series of elements, Boulez said: 'We threw our souls and our bodies into mathematics, tossing ciphers.'

IRCAM/Georges Meguerditchian

The theatre of life

Combining instrumental, environmental and electronic music with mirrors and laser projections, reminds us in many ways of antique rituals. *Mousike*, in Ancient Greek, meant a synthesis of music, dance and drama, and its god of death and fertility, Dionysus, the son of Zeus and a mortal woman, was unique in the Pantheon.

The Greeks liked to think that the gods could give life to their statues. Their spirits entered Western civilization via Venice and, within half a century, Europe was flooded with stage gadgets and automates.

In Rome, Bernini became the most brilliant and versatile decorator of his age. The papal architect and sculptor built an opera house, painted the sets, carved the statues, invented the stage machinery, and even composed music! Angels, devils and light flashes captivated the excited crowds. In baroque staging techniques, visual and musical expression were intimately linked, almost as in cinema.

Ballet performances were particularly admired for the devices and moving mechanical props used to create special effects. Italians introduced them at Versailles, where geometrical dance patterns were developed. King Louis XIV himself took active part and acquired his nickname, *le Roi Soleil*, from the headdress of sun-rays he wore in one performance. The dance Academy was founded to promote the highest standards, at about the same time as the science Academy.

OLYMPIC THEATRE STAGE, DESIGNED BY VINCENZO SCAMOZZI, AFTER PALLADIO, SEVENTEENTH CENTURY
Throughout the history of the performing arts, scientists, engineers and architects have joined forces with painters, dancers and musicians to create performances ahead of the fashion of their day. During Antiquity, illusionistic scenery was already produced. This stage is remarkable for its perspectival Renaissance design.
Alinari-Giraudon

Heavy costumes restrained leg movements and choreography became academic. After nearly three hundred years of regression, the great scenic Renaissance tradition of combining art with science and technology was revived by the Russian impresario Sergei Diaghilev (1872–1929).

Such an integral approach to sound, movement and scenery dominates modern collaborations by John Cage, composer and painter, Merce Cunningham, choreographer, and the visual artist, Robert Rauschenberg. The latter was especially interested in bringing artists and scientists together through his Experiments in Art and Technology (EAT) which united almost six thousand creators.

John Cage, the founder of the Happening, was instrumental in breaking down the barriers between art, science and life. He considered 'the sounds produced by the musical cultures of the universe' to be more interesting than those of the concert hall.

LA CRÉATION DU MONDE, STAGE MODEL, FERNAND LÉGER, 1923
Léger, a friend of Le Corbusier and a sincere admirer of machine aesthetics, created huge geometrical compositions to glorify the modern utopia of science.
Dansmuseet, Stockholm

Ballet was one of the earliest arts to have its own Academy, yet it was also the last to acquire a capacity for regular progression.

Until recently, there was no technical way of recording dance steps. For example, the recreation of *The Rite of Spring* set to Stravinsky's music – a milestone in the history of dance – used to be a choreographer's nightmare. Today, using sensors, details of movements can be recorded in all dimensions.

Attempts at recording motion through the mechanical manipulation of light and machines were numerous. At the end of the nineteenth century, kinetic devices were popular attractions in Paris, London and Chicago. The Englishman Eadweard Muybridge(1830–1904) advanced photography around the same time as Étienne-Jules Marey (born and deceased in the same years as Muybridge) in France. They both used the photograph as a tool for demonstrating the mechanics of human and animal locomotion.

Marey, the inventor of the instrument used to measure blood pressure, developed motion photography into a true art form. His chronophotography captured different phases of a single movement on the same photographic plate. In order to obtain articulate schemes, Marey asked a walker to: 'wear an entirely black costume, covered with narrow shiny metal strips, aligned along the legs and arms, indicating the orientation of the bones.' The photographic result was magical.

Both Muybridge and Marey were soon eclipsed by the Lumière brothers, who developed the moving picture camera: they took the pictures out of the box, so to speak, by projecting them on to a screen; their cinematograph functioned both as camera and projector. Thomas Edison, however, by his multiple inventions, was probably the greatest contributor to cinema and the performing arts.

STROBOSCOPIC DISK, J. A. PLATEAU. WHEEL OF LIFE, W. G. HORNER, NINETEENTH CENTURY
Attempts to mimic motion were numerous. Moving images projected on to large circular screens attracted crowds watching from the centre. Cinema was an immediate success, outdating all previous inventions.
Martin Kemp, The Science of Art. *Yale University Press. Pelican History of Art*

In the year that Roentgen discovered X-rays (1895), cinematographic devices appeared in different countries. The fundamental novelty of all these inventions lay in the interchangeability of space and time. Battles for patents were raging as moving pictures stole part of the glamour from the stage. By 1912, London had four hundred cinemas, and some movie theaters in New York could seat as many as 2,000 people.

Cinema continued to progress significantly with the addition of sound and colour in the 1930s. The introduction of computer graphics encouraged creativity in animation and ever more special effects. Technology, however, often plays second fiddle to art in cinema: the human dimension still dominates.

Film-making is now a widely recognized art form that would amaze some of its pioneers, who sought to use technologies and tricks to understand body mechanics, or merely to communicate . . . laughter and gags.

HORSE IN MOTION, EADWEARD MUYBRIDGE, 1878
California Governor, Leland Stanford, argued with a friend about the exact stride of a horse during gallop. Muybridge won a $25,000 prize for documenting the answer: all hoofs are off the ground simultaneously (see row above, in the middle). He experimented with up to twenty-four cameras used simultaneously, and published a book on animal locomotion containing over 20,000 sequences of movements.
© Hulton Getty/Fotogram-Stone Images, Paris

La Fée Électricité, Raoul Dufy, 1930

The French painter Dufy had a great interest in industrial design. The above 250-panel work portrays over a hundred scientists whose names are linked to the invention of electricity. Presented in chronological order and allegorical sequence, they are shown, from Thales to Einstein, puzzling out the mystery of science.

Musée d'Art Moderne de la Ville de Paris.

Although the silver screens of old have been transformed by all-round effects into all-encompassing multi-dimensional environments, cinema's real competition comes from a relatively unimposing box.

The invention of television was another one of those unexpected by-products of research that revolutionized art, science – and much else. In trying to increase the resolution of the microscope around 1930, Russian-born Vladimir Zworykin invented the electronic tube, which was to become the precursor of television.

He obtained a patent for colour television shortly afterwards and eventually succeeded, some ten years later, in improving the microscope's magnifying power. Within a few years, public television was under way. The Canadian sociologist and media specialist Marshall McLuhan said of television that it is 'light through not light on. . . . The image so formed has the quality of sculpture and icon rather than of picture.'

Television was long considered a manipulative instrument and still is, by some. Video exploits the quality of instantaneousness, as does television, but further liberates us from the constraint of time through recording, thus providing an additional degree of freedom.

The Korean artist Nam June Paik had already predicted thirty years ago that 'some day artists will work with capacitors, resistors and semi-conductors [components of television] as they work today with brushes, violins and junk'.

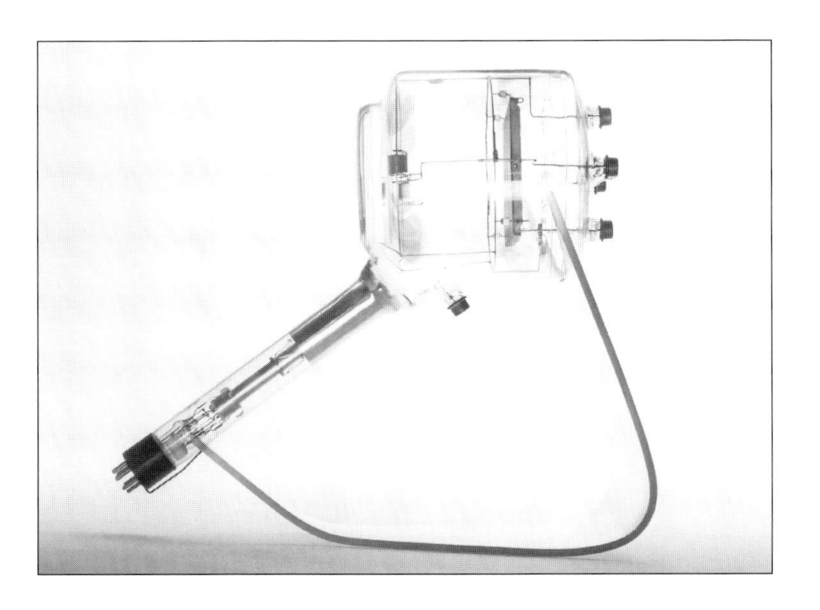

ICONOSCOPE
Light has become the vehicle of information. This tool, a precursor of the basic component of television, uses a material that reacts to light by emitting electrons – allowing electronic transmission of pictures.

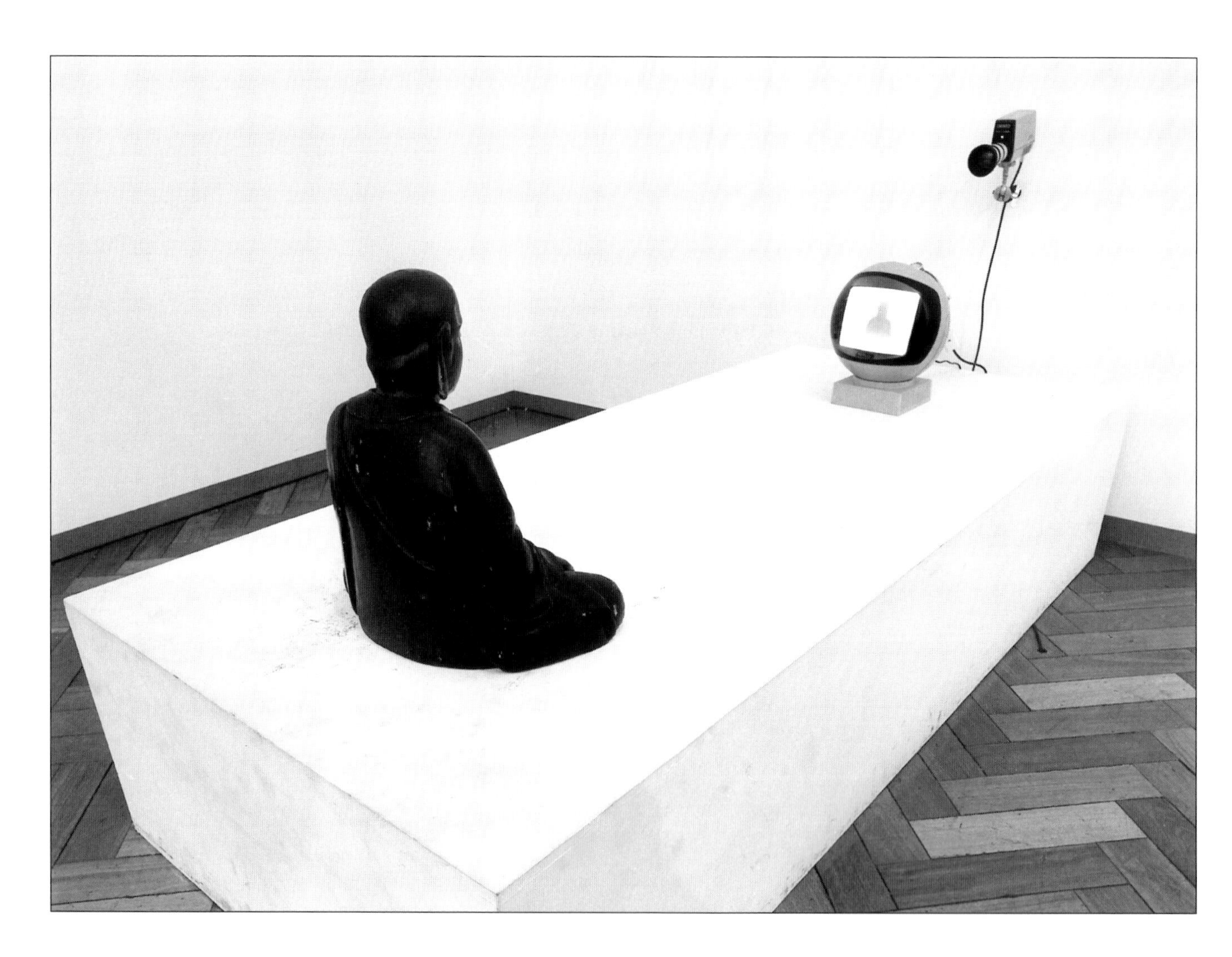

TV Buddha, Nam June Paik, 1974

This video installation with its meditating statue comments on technology's ubiquitous impact on society. Modern communication tools are often appropriated by artists.

Stedelijk Museum Amsterdam, the Netherlands

After the electronic transition began with broadcasting, art boldly moved from the mechanical to the electronic age, in an unparalleled marriage with science. The next step would be the recording of sound and light initially achieved by using magnetic tapes.

Then, the invention of the laser (light amplification by stimulated emission of radiation) allowed for radical progress in recording. One of its revolutionary effects was that music as well as images, converted into light signals, could be stored on discs and reproduced with unprecedented fidelity.

The video and audio discs that we know as compact discs (CDs) function on the same principle: below the plastic surface, billions of tiny pits encoding music or images can be deciphered by a laser beam. In 1917, Einstein had already predicted the invention of the laser. Today, lasers have become most versatile, and are particularly potent when transmitted through fibre-optic glass. They are manufactured for practical applications in medicine or telecommunications; daring artists have seized upon them, too.

The potential of the laser has been enhanced by yet another new technique: holography, which was discovered somewhat unintentionally by the Hungarian physicist Dennis Gabor, while trying to improve the quality of electronic images. A wide array of holographic applications is being investigated, ranging from aircraft cockpit scanners to bar codes.

As holograms open up a new aesthetic domain, by stimulating spatial imagination, one might recall that the Greek word *esthetike* originally meant – perhaps prophetically – what we call 'sensation'. The quest to translate visual reality into new perceptions and performance genres may be more than just an illusion; holographic television images are a short distance away. Meanwhile, three-dimensional reality has invaded daily life through computer graphics.

Will the eighth art form be of the virtual type?

An art and science symbiosis

In seeking clues to the secrets of the universe, scientists send waves into the cosmos and frequently bestow odd, affectionate names – gluon, quark, charm – on their fascinating finds. Artists adopt these poetic concepts and translate them into visible forms.

Some artists use new technologies to nurture the illusion of a co-evolution between man and machine. Others, on the contrary, avoid technology, which remains expensive and sometimes complex to access. Further application of science is directly linked to the ability to cost-effectively integrate fast-evolving technologies. Hi-tech media are at the creator's fingertips but, at the same time, the artist's fundamental copyright can be easily overlooked: images can instantly be manipulated on a planetary scale.

Against this backdrop of 'techno-maternity', as the writer René Berger calls it, the race for more information and innovative art forms is more competitive than ever. Yet, creativity in art and science no longer needs to be a specialist domain. Today's most challenging competitor is the adolescent surfing on the Internet, caring little about who is an artist or a scientist.

It is easy to escape the structured mode, by zapping through a welter of sound and images. This mental zigzagging through masses of information appeals increasingly to both parts of the brain. The use of the left side, more concerned with logic, may become slightly displaced by a flexible and fragmented mode of thinking. The brain's right side, seat of the imagination, may be more developed in a society which quite simply has more time to hand.

Some forms of expression, such as painting, seem to be losing ground, while others flourish. Science is probably among the culprits. Technology has transformed the fundamental relationship between artist and art form: the painter to the hand and eye; the musician to the ear. As such, the very nature of communication between performer and public is modified.

When composed with the aid of computers, art becomes a pluridisciplinary affair: painting has access to millions of hues;

photography uses numerical images that can be transformed at will; and video animates these images.

Today, laser technology, holograms and computer synchronization used in multi-media performances, appeal to artists and scientists as well as to a broad public. The transdisciplinary culture has come full circle. There are nearly as many styles as there are creators – whether innovators like Cézanne or Einstein, or synthesizers like Raphael or Edison.

During the past few decades, traditional boundaries between the arts and sciences have gradually eroded, and *techne* has found once again her antique spiritual dimension. The gods of aesthetics and logic have returned to Olympia, making room for poetry and curiosity on a daily basis.

KEO, The Archeological Bird of the Future,
Jean-Marc Philippe, 1997
This strange bird – a passive satellite containing messages from artists, scientists, philosophers – is intended for our distant descendants. It will be 'injected' into an appropriate space orbit in 2001, and will spontaneously return to earth 50,000 years later.
Infographics R. Locicero. Aérospatiale, Sup-Aero, CEA
www.keo.org

Bibliography

General bibliography

ARTEC. First and Second Biennial Exhibitions Catalogue. Nagoya, 1989 and 1991.

CALVESI, M. Biennial Exhibition Catalogue, *Art and Science.* Venice, 1986.

CHANGEUX, J.-P. *Raison et plaisir.* Paris, Odile Jacob, 1994.

CLAIR, J.; CHANGEUX, J.-P. Exhibition Catalogue, *L'âme au corps : art et sciences 1793–1993,* Grand Palais. Paris, Réunion des musées nationaux – Gallimard, 1993–4.

COOK, T. *The Curves of Life.* New York, Dover, 1979.

GRAUBARD, S. R. (ed.). Art and Science. *The Daedalus Journal of the American Academy of Arts and Sciences.* Lanham, Maryland, University Press of America, 1986.

HARTAL, P. Z. *The Brush and the Compass. The Interface Dynamics of Art and Science.* Lanham, Maryland, University Press of America, 1988.

HASKELL, F. *Art as a Prophecy. History and its Images.* New Haven, Yale University Press, 1995, pp. 389–430.

HOFSTADTER, D. R. *Gödel, Escher, Bach: An Eternal Golden Braid.* New York, Basic Books, 1980.

KAPRAFF, J. *Connections, the Geometric Bridge between Art and Science.* New York, McGraw-Hill Inc., 1991.

KEPES, G. *The New Landscape in Art and Science.* Chicago, Theobald and Co., 1956.

LESHAN, L.; MARGENAU, H. *Einstein's Space and Van Gogh's Sky.* New York, Collier, 1982.

LEVY, E. K.; SICHEL, B. M. (eds.). Contemporary Art and the Genetic Code. *Art Journal,* New York, College Art Association, Vol. 55, No. 1, 1996.

MERLEAU-PONTY, M. *L'œil et l'esprit.* Paris, Gallimard, 1964.

NAGUY, L. *Vision in Motion.* Chicago, 1947.

POLLOCK, M. (ed.). *Common Denominators in Art and Science.* Aberdeen University Press, 1983.

POPPER, F. Exhibition Catalogue, *Électra: l'électricité et l'électronique dans l'art au XXe siècle.* Musée d'Art Moderne de la Ville de Paris, 1983.

ROBIN, H. *The Scientific Image.* New York, Freeman, 1992.

ROOT-BERNSTEIN, R. The Visual Arts and Sciences. *Transaction.* Philadelphia, The American Philosophical Society, Vol. 75/6, 1985, pp. 50–68.

SHLAIN, L. *Art and Physics.* New York, William Morrow and Co., 1991.

SMITH, C. S. *A Search for Structure. Selected Essays on Science, Art and History.* Cambridge, Massachusetts, MIT Press, 1981.

THOMPSON D'ARCY, W. *On Growth and Form.* Cambridge University Press, 1961.

TOPPER, D. R. Historical Perspectives on the Arts, Sciences and Technology. *Leonardo,* Cambridge, Massachusetts, MIT Press, Vol. 18, No. 1, 1985.

VALÉRY, P. *Selected Writings of Paul Valéry.* New York, W. W. Norton, 1991.

VITZ, P. C.; GLIMCHER, A. B. *Modern Art and Modern Science.* New York, Praeger, 1984.

WECHSLER, J. *On Aesthetics in Science.* Cambridge, Massachusetts, MIT Press, 1981.

WIJERS, L.; PIJNAPPEL, J. *Art meets Science and Spirituality.* New York, St Martin's Press, 1990. (Art and Design.)

Art history

ARNHEIM, R. *Visual Thinking.* Berkeley, University of California Press, 1970.

BECK, J.; DALEY, M. *Art Restoration: the Culture, the Business and the Scandal.* London, John Murray, 1993.

COOK, E. T.; WEDDERBURN, A. (eds.). *The Works of John Ruskin.* London/New York, 1903–12.

FOCILLON, H.; KUBLER, G. (transl.). *The Life of Forms in Art.* Zone Books, 1992.

GABLIK, S. *Progress in Art.* London, Thames and Hudson, 1976.

GOETHE, J. W. In J. VON GEARY (ed.), *Essays on Art and Literature.* Princeton University Press, 1994.

GOMBRICH, E. H. *Art and Illusion.* New York, Pantheon Books, 1972.

GREENBERG, C. *Art and Culture.* Boston, Beacon Press, 1961.

HUYGHE, R. *Formes et forces de l'atome à Rembrandt.* Paris, Flammarion, 1971.

JANSON, H. W. *History of Art.* New York, Harry N. Abrams, 1991.

LOGAN, C.; LIEDTKE, W. A.; VAN SONNENBURG, H. Exhibition Catalogue, *Rembrandt/Not Rembrandt: Aspects in Connoisseurship.* The Metropolitan Museum of Art, New York, 1995.

READ, H. E. *The Philosophy of Modern Art.* London, Faber and Faber, 1982.

—. *Art and Alienation: the Role of the Artist in Society.* New York, Horizon, 1967.

RENFREW, C.; BAHN, P. *Archaeology, Theories, Methods and Practice.* London, Thames and Hudson, 1991.

VASARI, G.; HINDS, A. B. (transl.). In W. GAUNT (ed.), *Lives of the Most Eminent Painters, Sculptors and Architects.* New York, Dutton, 1963.

WOOD, C. S. PANOFSKY, E. *Meaning in the Visual Arts.* Woodstock, Overlook Press, 1974.

—. (transl.). *Perspective as Symbolic Form.* Zone Books, 1997.

History of science and ideas

BOORSTIN, D. J. *The Discoverers.* New York, Vintage Books, 1983.

BURKE, J. *The Day the Universe Changed.* London and Boston, Little, Brown and Co., 1985. BBC television series.

CANGUILHEM, G.; DELAPORTE, F. (eds.). *A Vital Rationalist: Selected Writings.* Zone Books, 1994.

CLARK, R. W. *Einstein: The Life and Times.* New York, Avon, 1971.

DIDEROT, D.; ALEMBERT, J. LE ROND D'. *A Pictorial Encyclopedia of Trades and Industry.* New York, Dover Publications Inc., 1993.

HANCOCK, G. *Fingerprints of the Gods: A Quest for the Beginning and the End.* New York, Crown, 1995.

KOYRÉ, A. *Metaphysics and Measurement: Essays in the Scientific Revolution.* Cambridge, Massachusetts, Harvard University Press, 1968.

KUHN, T. *The Structure of Scientific Revolutions.* University of Chicago Press, 1996.

LANDES, D. S. *Revolution in Time: Clocks and the Making of the Modern World.* Cambridge, Massachusetts, Harvard University Press, 1983.

NEEDHAM, J. *Science and Civilization in China.* Cambridge University Press, 1954.

NEUGEBAUER, O. *The Exact Sciences in Antiquity.* New York, Dover Publications, 1957.

POPPER, K. *Objective Knowledge. An evolutionary approach.* Oxford, 1972.

RONAN, C. *The Cambridge Illustrated History of the World's Science.* Twickenham, Middlesex, Newnes Books, 1983.

RUSSEL, B. *History of Western Philosophy.* London, Unwin Paperbacks, 1982.

SEAMON, D.; ZAJONC, A. (eds.). *Goethe's Way of Science: A Phenomenology of Nature.* State University of New York, 1998.

VAN DOREN, C. *A History of Knowledge.* New York, Ballantine, 1992.

Creativity and cognitive sciences

BOORSTIN, D. J. *The Creators.* New York, Vintage Books, 1993.

COLLINS, A.; SMITH, E. E. (eds.). *Readings in Cognitive Science: a Perspective from Psychology and Artificial Intelligence.* San Mateo, California, Morgan Kaufmann Publishers Inc., 1988.

EDWARDS, B. *Drawing on the Artist Within.* New York, Simon and Schuster, 1986.

EINSTEIN, A. Autobiographical Notes. In P. A. SCHILPP (ed.), *Albert Einstein: Philosopher-Scientist.* Evanston, Illinois, The Library of Living Philosophers, 1949.

GARDNER, H. *Art, Mind and Brain. A Cognitive Approach to Creativity.* New York, Basic Books, 1982.

—. *Creating Minds.* New York, Basic Books, 1993.

—. *Multiple Intelligences.* New York, Basic Books, 1993.

HADAMARD, J. *The Psychology of Invention in the Mathematical Field.* New York, Dover, 1954.

JUNG, C. G. *Man and his Symbols.* London, Aldus, Jupiter, 1979.

JUNG, R. *Psychiatrie der Gegenwart.* Berlin-Heidelberg-New York, Spring, 1980.

KOESTLER, A. *The Act of Creation.* London, Hutchison, 1976.

MILLER, A. I. *Insights of Genius. Imagery and Creativity in Science and Art.* New York, Springer-Verlag, 1996.

PIAGET J. In P. MUSSEN (ed.), *Piaget's theory. Manual of Child Psychology.* New York, Wiley, 1983.

POINCARÉ, H. *Science and Hypothesis.* New York, Dover, 1952.

RAMON Y CAJÁL, S. *Recollections of my Life.* Philadelphia, The American Philosophical Society, 1937.

SOLSO, R. L. *Cognition and the Visual Arts.* Cambridge, Massachusetts, MIT Press, 1994.

VAN T'HOFF, J. H. Imagination in Science. *Mol. Biol. Biochem. Biophys.*, 1967.

WEISBERG, R. *Creativity: Genius and other Myths.* New York, Freeman, 1986.

WOTIZ, J. H; RUDOFSKY, S. Kékulé's Dream: Fact or Fiction? *Chemistry in Britain,* 1954.

ZEKI, S. *A Vision of the Brain.* Oxford, Blackwell, 1993.

SCIENCE IN ARCHITECTURE

FRAMPTON, K.; CAVA, J. (eds.). *The Poetics of Construction in the Nineteenth and Twentieth Century.* Cambridge, Massachusetts, MIT Press, 1996.

GALISON, P. *Aufbau/Bauhaus: Logical Positivism and Architectural Modernism.* Critical Inquiry, 1990.

GIEDION, S. *Space, Time and Architecture.* Cambridge, Massachusetts, Harvard University Press, 1967.

KOSTOFF, S. *History of Architecture.* New York, Oxford University Press, 1995.

LE CORBUSIER. *Modulor.* Cambridge, Massachusetts, MIT Press, 1968.

LONG, DE G.; BROWNLEE, D. B. *Louis I. Kahn: In the Realm of Architecture.* New York, Universe Publishing Inc., 1997.

PANOFSKY, E. *Gothic Architecture and Scholasticism.* Latrobe, Pennsylvania, Archabbey Press, 1951.

PEVSNER, N. *The Sources of Modern Architecture and Design.* Oxford, Oxford University Press, 1977.

PICON, A. (ed.). Exhibition Catalogue, *L'art de l'ingénieur. Constructeur, entrepreneur, inventeur.* Paris, Éditions du Centre Pompidou–Le Moniteur, 1997.

SCHOLFIELD, P. H. *The Theory of Proportion in Architecture.* Cambridge University Press, 1958.

VIOLLET-LE-DUC, E.; In M. F. HEARN (ed.), *Architectural Theory: Readings and Commentary.* Cambridge, Massachusetts, MIT Press, 1990.

VITRUVIUS. In M. H. MORGAN (ed.), *The Ten Books on Architecture.* New York, Dover, 1960.

DECORATION: A PATH TO HI-TECH

CHRISTIE, A. H. *Pattern Design.* New York, Dover, 1969.

CRITCHLOW, K. *Islamic Patterns; an Analytical and Cosmological Approach.* London, Thames and Hudson, 1989.

ENGEL, P. *Folding the Universe; Origami from Angelfish to Men.* New York, Vintage Books, 1989.

GOMBRICH, E. H. *The Sense of Order, a Study in the Psychology of the Decorative Arts.* London, Phaidon, 1994.

GRUNBAUM, S. *Tiling and Patterns.* New York, Freeman and Co., 1987.

HOGARTH, W. *The Analysis of Beauty, written with a View of fixing the fluctuating Ideas of Taste.* London, 1753.

MANDELBROT, B. *The Fractal Geometry of Nature.* New York, W. H. Freeman, 1983.

MCCARTY, C. *Information Art. Diagramming Microchips.* New York, The Museum of Modern Art, 1990.

OSBORNE, H. (ed.). *The Oxford Companion to the Decorative Arts.* Oxford, Clarendon Press, 1975.

PASTEUR, L. *Les forces cosmiques asymétriques.* Unpublished manuscript. Paris, Institut Pasteur.

RHODES, L. *Science within Art.* The Cleveland Museum of Art, 1980.

SMITH, C. S. *A History of Metallography.* Chicago University Press, 1960.

—. *Aspects of Art and Science.* Cambridge, Massachusetts, MIT Press, 1978.

STEVENS, P. *Handbook of Regular Patterns: an Introduction to Symmetry in Two Dimensions.* Cambridge, Massachusetts, MIT Press, 1981.

WEYL, H. *Symmetry.* Princeton University Press, 1952.

ZERWICK, C. *A Short History of Glass.* The Corning Museum of Glass. New York, Abrams, 1990.

PAINTING AND COGNITION

ALBERS, J. *Interaction of Colors.* New Haven, Yale University Press, 1975.

ALBERTI. *On Painting.* New Haven, Yale University Press, 1966.

APOLLINAIRE, G.; SULEIMAN, S. (transl.). In I. BREUNIG (ed.), *On Art: Essays and Reviews (1902–1918).* Da Capo, 1988.

ARNHEIM, R. *The Genesis of a Painting: Picasso's Guernica.* Berkeley, University of California Press, 1962.

BALTRUŠAITIS, J. *Anamorphic Art.* Cambridge University Press, 1977.

BERNARD, E.; GASQUET, J. (eds.). *Conversations avec Cézanne.* Paris, Macula, 1978.

BRAQUE, G. *Le jour et la nuit.* Paris, Gallimard, 1952. (Cahiers.)

BRUSATIN, M. *A History of Colors.* Boston, Shambhala Publications, Inc., 1991.

BURWICK, F.; DE GRUYTER, W. (eds.). *The Damnation of Newton: Goethe's Color Theory and Romantic Perception.* Dusseldorf, Walter Verlag, 1986.

CACHIN, F. *Seurat. Le rêve de l'art-science.* Paris, Gallimard, 1991. (Découvertes.)

CÉZANNE, P. In I. CAHN, H. LOYRETTE, J. J. RISHEL and F. CACHIN (eds.), *Cézanne.* New York, Harry N. Abrams, 1996.

CLAIR, J. *Éloge du visible.* Paris, Gallimard, 1996.

—. Psychanalyse et symbolisme. Catalogue de l'exposition *Paradis perdus : l'Europe symboliste.* Montreal, Musée des Beaux-Arts/Paris, Flammarion, 1995.

CLOUZOT, H.-G. *Le mystère Picasso.* Film, 1955.

COMAR, P. *La perspective en jeu. Les dessous de l'image.* Paris, Gallimard, 1992. (Découvertes.)

—. *Les images du corps.* Paris, Gallimard, 1993. (Découvertes.)

CONSTABLE, J. *Memoirs of the Life of John Constable.* London, Jonathan Mayne, 1951.

DUPARC, F. J.; WHEELOCK, A. K. JR. (eds.). Exhibition Catalogue, *Johannes Vermeer.* Washington, National Gallery of Art/The Hague, Mauritshuis. Zwolle, Waanders Publishers, 1996.

EDGERTON, S. Y. JR. *The Heritage of Giotto's Geometry: Art and Science on the Eve of the Scientific Revolution.* Ithaca, New York, Cornell University Press, 1991.

FLAM, J. Exhibition Catalogue, *De Cézanne à Matisse. Chefs-d'œuvre de la fondation Barnes.* Paris, Gallimard Électa/Réunion des musées nationaux, 1993, pp. 274–91.

FREUD, S. In J. STRACHEY (ed.), *Civilization and its Discontents.* New York, W. W. Norton and Company, 1989.

GINZBURG, C. *Enquête sur Piero della Francesca.* Paris, Flammarion, 1983.

GLEIZES, A. *Art et religion, art et science, art et production.* Chambéry, Présence, 1970.

HELMHOLTZ, H. *Handbuch der Physiologisschen Optik.* Hamburg–Leipzig, Voss, 1867.

HENDERSON, L. *The Fourth Dimension and Non-Euclidian Geometry in Modern Art.* Princeton University Press, 1984.

HENRY, C. *Paul Signac.* Paris, Cahiers de l'Étoile, 1931.

HULTEN, P. (ed.). *Futurismo & futurismi.* Milan, Fabbri, 1986.

KAHNWEILER, D. H. *Juan Gris, sa vie, son œuvre.* Paris, Gallimard, 1946.

KANDINSKY, W. *Concerning the Spiritual Art.* New York, Dover, 1977.

KEMP, M. *The Science of Art: Optical Themes in Western Art from Brunelleschi to Seurat.* New Haven, Yale University Press, 1992.

KLEE P. In H. DAY (ed.), *On Modern Art Advances in Intrinsic Motivation and Aesthetics.* New York, Plenum Press, 1981.

LORAN, E. *Cézanne's Composition: Analysis of his Form with Diagrams and Photographs of his Motifs.* Berkeley, University of California Press, 1947.

MALEVITCH, K. *Le miroir suprématiste.* Lausanne, L'âge d'homme, 1977.

MATISSE, H. *Écrits et propos sur l'art.* Paris, Hermann, 1972.

MOLNAR, F. About the Role of Visual Exploration in Aesthetics. In H. DAY (ed.), *Advances in Intrinsic Motivation and Aesthetics.* New York, Plenum Press, 1981.

MONDRIAN, P. In H. B. CHIPP (ed.), *Plastic Art and Pure Plastic Art. Theories of Modern Art.* Berkeley, University of California Press, 1968.

PENROSE, R. *Max Ernst's Celebes.* University of Newcastle-upon-Tyne, 1972.

PRINZHORN, H.; BROCKDORFF, E. Von (transl.). In J. L. FOY (ed.), *Artistry of the Mentally Ill: A Contribution to the Psychology and Psychopathology of Configuration.* New York, Springer-Verlag, 1995.

SIGNAC, P. *Neo-impressionism.* Concord, Massachusetts, Paul & Company Publishers Consortium, Inc., 1990.

WADDINGTON, C. H. *Behind Appearances, a Study of the Relations between Painting and the Natural Sciences this Century.* Cambridge, Massachusetts, MIT Press, 1970.

THE LANGUAGE OF GRAPHIC DESIGN

ANDERSON, D. *The Art of Written Forms. The Theory and Practice of Calligraphy.* New York, Holt, Rinehart and Winston, 1969.

BAUMUNK, B. M. Exhibition Catalogue, *Darwin und Darwinismus.* Deutsches Hygiene Museum Dresde. Berlin, Akademie Verlag, 1994.

DÜRER, A. *Underweisung der Messung.* New York, Albaris Books, 1977.

FORD, B. J. *A History of Scientific Illustration.* New York, Oxford University Press, 1992.

FRIDLUND, A. J. *Human Facial Expression: an Evolutionary View.* London, Academic Press, 1994.

HOOKE, R. In R. T. GUNTHER (ed.), *Micrographia.* Early Science in Oxford series, Vol. 13. Oxford University Press, 1938.

JEAN, G. *Writing. The Story of Alphabets and Scripts.* New York, Abrams Discoveries, 1987.

MCLUHAN, M. *The Gutenberg Galaxy: The Making of Typographic Man.* University of Toronto, 1962.

—. *Understanding Media: The Extensions of Man.* New American Library, 1964.

MEEHAN, B. *The Book of Kells.* London, Thames and Hudson, 1995.

MEGGS, P. B. *A History of Graphic Design.* New York, Van Nostrand Reinhold, 1992.

RODARI, F. Exhibition Catalogue, *Anatomie de la couleur. L'invention de l'estampe en couleurs.* Paris, Bibliothèque Nationale de France/Lausanne, Musée Olympique, 1996.

ROGER, J.; BONNEFOI, L. S. (transl.). *Buffon.* Ithaca, Cornell University Press, 1997.

ROSENBLUM, N. *A World History of Photography.* New York, Abbeville Press, Inc., 1997.

SACHSSE, R. *Blossfeldt.* Cologne, Taschen, 1996.

SCHWARTZ, L.; SCHWARTZ, L. *The Computer Artist's Handbook.* New York, Norton, 1992.

TECHNIQUE AND THE PERFORMING ARTS

ABRAHAM, G. *The Concise Oxford History of Music.* New York, Oxford University Press, 1991.

BRAUN, M. *Picturing Time: The Work of Étienne-Jules Marey.* University of Chicago Press, 1994.

COPE, D. *Computers and Musical Style.* Berkeley, University of California Press, 1991.

HELMHOLTZ, H. *On the Sensations of Tone.* New York, Dover, 1954.

JUNG, D. Holographic Space: a Historical View and some Personal Experiences. *Leonardo.* Cambridge, Massachusetts, MIT Press, Vol. 22, Nos. 3–4. 1989, pp. 331–6.

LEVENSON, T. *Measure for Measure, a Musical History of Science.* New York, Simon and Schuster, 1994.

MUYBRIDGE, E. *Complete Human and Animal Locomotion.* New York, Dover Publications, 1979.

POPPER, F. *Art of the Electronic Age.* London, Thames and Hudson, 1993.

RESTANY, P. *Catherine Ikam: le grand jeu de la vidéo.* Paris, Maeght, 1991.

SACCONE, S. *The Secrets of Stradivari.* Cremona, Libraria del Convegno, 1972.

THOMPSON, S. P. *Light, Visible and Invisible.* New York, Macmillan, 1987.

XENAKIS, I. *Formalized music.* Bloomington, Indiana University, 1971.

—. *Diatope,* plastic and musical composition, 1978.

ZENTRUM FÜR KUNST UND MEDIENTECHNOLOGIE, multi-media activity programme. Karlsruhe, http://wwwzkm.de

Index of names

Index of subjects

Credits

This book would not have been possible without the generosity of those who lent their iconographic documents as a courtesy:

Artists, authors or their heirs

Chermann J.-C.,100. Comar Philippe, 48, 70. Ikam Catherine, 18. Jung Dieter, 211. Karavan Dani, 71. Kemp Martin, 154, 224. Kriki, 166. Leviant Isia, 167. Libeskind Daniel, 95. Mandelbrot Benoit, 100. Marshack Alexander, 40. Molnar Vera, 169. Noguchi Mikiko, 56, 57, 67, 94. Parker Bill, 23. Philippe Jean-Marc, 232. Schulmann Jean-Louis, 83. Schwartz Lillian, 12, 207. Shaw Jeffrey, 22. Snelson Kenneth, 97. Starck Philippe, 93. Strosberg Serge, 33, 34, 51.

Museums

Ägyptisches Museum im Bodemuseum, Berlin, 176. Bayerischen Staatsgemäldesammlungen, Munich, 156. British Museum, London, 42. Cincinnati Art Museum, 191. Cleveland Museum of Art, 107, 123, 125, 137. Corning Museum of Glass, 108, 109, 113. Dansmuseet, Stockholm, 223. Egyptian Museum in Cairo, 108. Freer Gallery of Art, Washington D. C., 124. Galleria Borghese, Rome, 150. Galleria Nazionale delle Marche, Urbino, 148. Gallerie dell'Accademia, Venice, 18. George Eastman House, Rochester, 87, 134, 198, 199. Germanisches Nationalmuseum, Nuremberg, 115. Graphisches Sammlung Albertina, Vienna, 160. Kunsthistorisches Museum, Vienna, 210. Mauritshuis, The Hague, the Netherlands, 152. Metropolitan Museum of Art, New York, 25, 27, 139. Musée Condé, Chantilly, 181, 213. Musée d'Art Moderne de la Ville de Paris, 226. Musée de l'Assistance Publique, Paris, 81. Musée des Antiquités Nationales de Saint-Germain-en-Laye, 40. Musée des Arts et Métiers-CNAM, Paris, 69, 103, 110, 218, 227. Musée des Instruments de Musique de Bruxelles, 216. Musée d'Orsay, 156, 169. Musée du Louvre, 44, 154. Musée National d'Art Moderne-Centre de Création Industrielle-Centre Georges Pompidou, Paris, 22, 86, 160, 165. Musée National des Arts Asiatiques-Guimet, 142. Musées Royaux des Beaux-Arts de Bruxelles, 75. Museo Archeologico Nazionale, Naples, 140. Museo e Gallerie Nazionali di Capodimonte, Naples, 102. Museum für Völkerkunde, 60, 62. Museum of Fine Arts, Boston, 119, 175, 183, 189. Museum of Modern Art, New York, 160, 203. National Gallery, London, 152. National Gallery of Art, Washington D. C., 54, 139. National Maritime Museum, Greenwich, London, 69, 79. National Museums of Scotland, 121, 153. Neue Pinakothek, Munich, 156. Observatoire de Paris, 73. Ontario Science Center, 182. Philadelphia Museum of Art, 143, 163. Potteries Museum, Stoke-on-Trent, 122. Rheinishes Landesmuseum, Bonn, 53. Rijksmuseum, Amsterdam, 152. Science Museum, London, 66, 94, 95, 104, 120, 127, 192. Staatliche Museen, Berlin, 26, 60, 62, 116, 117, 176. Stedelijk Museum, Amsterdam, 228. Tate Gallery, London, 28, 63, 155. Terra Museum of American Art, Chicago, 30. Topkapi Palace Museum, Istanbul, 105. University of Pennsylvania Museum, Philadelphia, 42, 117, 174. Vatican Museums, 149. Victoria & Albert Museum, London, 114, 121, 152, 176, 214, 215. Vorderasiatisches Museum, Berlin, 116.

Photographers, picture libraries and agencies, graphic designers

A.C.L., Brussels, 75. Alberto L., 82. Alinari, 76, 102, 148, 222. ALS Colordia, 176. Anderson, 76. Ann Ronan Picture Library, 83. Archivio Fotografico Soprintendenza Beni Artistici e Storici di Roma, 150. Artway, 58. Berizzi J. G., 40. Bildarchiv, ÖNB, 52, 58, 65, 128, 192. Blot Gérard, 156. bpk, bildarchiv preussischer kulturbesitz, Berlin, 26, 60, 62, 116, 117, 176. Carafelli Richard, 54. Charmet J.-L., 81, 172. Commerce Graphics Ltd. Inc., New Jersey, 201. Cordon Art B. V. Baarn, the Netherlands, 36, 65. Cussac F., 75. Delepelaire, 226. Dunonau F., 41. Emprin Nathalie C., 101. Faligot P., 110, 218. Geske Sigrid, 16, 135. Giraudon, 76, 102, 148, 181, 213, 222. Göken Klaus, 116. Hatala, 165. Hulton Getty-Fotogram-Stone Images, Paris, 158, 193, 225. Image Digitale, 83. Image Select, London, 83. Joly Thierry, 136. Klinger, 159. Laurentius J., 26, 117. Lavda Siegfried, 92. Levinthal Nina, 85. Locicero R., 232. Maeyaert Paul M. R., 145. MAG, 149. Meguerditchian Georges, 221. Migeat Philippe, 22. Minkkinen Arno Rafael, 101. Moore Chris, 131. Namuth Studio, 164. National Gallery Picture Library, London, 152. Photographic Services of The Metropolitan Museum of Art, New York, 25, 27, 139. Photothèque des Collections du Mnam/Cci/Centre Georges-Pompidou, Paris, 22, 86, 160, 165. Photothèque des Musées de la Ville de Paris, 226. Produzione Image & Communication SNC, 112. Réunion des Musées Nationaux, 44, 154, 156, 169, 184, 197. Rimond Patrick, 62. Roger D., 50. Roger-Viollet, Paris, 27, 59, 61, 184, 197. Schneider-Schütz W, 62. Schonburgh H. H., 63. Science & Society Picture Library, London, 66, 94, 95, 104, 127, 192. Sebert P., 166. Service photographique Bibliothèque Nationale de France, 19, 186, 188, 190, 212, 219. Seventh Square, 110, 218. Siener Joachim, 202. Studio Cnam, 69, 103, 218, 227. Studio Libeskind, 95. Tate Picture Library, 28, 63, 155. Termisien Hervé, 93. V&A Picture Library, London, 114, 121, 152, 176, 214, 215. Vigne Jean, 78. Walker Julie, 23. Zigrossi P., 149.

Corporations, foundations and organizations

Adagp, 22, 63, 86, 160, 163, 165, 168. Aérospatiale, 232. Apple Computer, Inc., 204. Association Curie and Joliot-Curie, 90. Capella Degli Scrovegni-Commune di Padova Settore Musei e biblioteche, 144. CEA, 232. DuPont de Nemours, 130. Edison National Historic Site, 209. Estate of Hans Namuth, 164. Hoechst AG, 92. IBM, 58. INSERM, 100. Institut Pasteur, 88, 111. IRCAM, 221. IRPA-KIK, Brussels, 75, 216. Ministerio per i Beni Culturali e Ambientali, Italy, 18. NTV Nippon Television Network Corporation, 31. Pollock-Krasner House and Study Center, East Hampton, New York, 164. Prinzhorn-Sammlung der Psychiatrischen Universitätsklinik, Heidelberg, 159. Rijksmuseum Stichting, Amsterdam, 152. Seminario, Treviso, 112. Smithsonian Institution, Washington D. C., 124. Stiftung Weimarer Klassik, 16, 135. Sup-Aero, 232. Terra Foundation for the Arts, 30. The Barnes Foundation, 157, 206. The Royal Collection-Her Majesty Queen Elizabeth II, 187. The Royal Society, London, 77. US Department of the Interior-National Park Service, 209. W & L. T., 131.

Publishers, libraries and archives

Biblioteca Nazionale Centrale, Florence, 74. Bibliothèque littéraire Jacques Doucet, Paris, 172. Bibliothèque Municipale de Versailles, 78. Bibliothèque Nationale de France, 19, 186, 188, 190, 197, 212. Bodleian Library, University of Oxford, 80. British Library, London, 68, 171, 177, 178. Cambridge University, 79, 179. Dover Publications Inc., 147. Éditions Gallimard 48, 70. Goethe-Schiller-Archiv, 16. HarperCollins Publishers, 205. Herzogin Anna Amalia Bibliothek, 135. Libraria del Convegno, Cremona, 216. Library of Congress, Washington D. C., 20, 64, 72, 195, 213. Louise M. Darling Biomedical Library, UCLA, 75, 89, 194. MIT Press, Cambridge, Massachusetts, 14, 99. National Library of Medicine, Bethesda, 143. New York Public Library, 188. Österreichische NationalBibliothek, Vienna, 52, 58, 65, 128, 192. Research Library of the Corning Museum of Glass, 113. Schott Muzik International, Mainz, 219. Scientific American, 104, 200. Stiftsbibliothek, St Gall, 55. University of California Press, 162. University Research Library, UCLA, 153. Wellcome Institute Library, London, 74, 120, 196. Württembergische Landesbibliothek, 202. Yale University Press, 154, 224.

We apologize to those we were unable to trace despite our efforts:
Janiszewsky Jerzy, 203. Lehner Mark, 43. Loran Erle, 162. Silcock A., 173, Stevens Peter, 14, 99.